STOCHASTIC PROCESSES AND OPTIMAL CONTROL

STOCHASTICS MONOGRAPHS

Theory and Applications of Stochastic Processes
A series of books edited by Mark Davis, Imperial College, London, UK

Volume 1
Contiguity and the Statistical Invariance Principle
P.E. Greenwood and A.N. Shiryayev

Volume 2
Malliavin Calculus for Processes with Jumps
K. Bichteler, J.B. Gravereaux and J. Jacod

Volume 3
White Noise Theory of Prediction, Filtering and Smoothing
G. Kallianpur and R.L. Karandikar

Volume 4
Structure Selection of Stochastic Dynamic Systems: The Information Criterion Approach
Sándor M. Veres

Volume 5
Applied Stochastic Analysis
Edited by M.H.A. Davis and R.J. Elliott

Volume 6
Nonlinear Stochastic Integrators, Equations and Flows
R.A. Carmona and D. Nualart

Volume 7
Stochastic Processes and Optimal Control
Edited by H.J. Engelbert, I. Karatzas and M. Röckner

STOCHASTIC PROCESSES AND OPTIMAL CONTROL

Edited by

Hans J. Engelbert
Friedrich-Schiller-Universität
Jena, Germany

Ioanis Karatzas
Columbia University
New York, USA

and

Michael Röckner
Rheinische Friedrich-Wilhelms-Universität
Bonn, Germany

GORDON AND BREACH SCIENCE PUBLISHERS
Switzerland • Australia • Belgium • France • Germany • Great Britain
India • Japan • Malaysia • Netherlands • Russia • Singapore • USA

Gordon and Breach Science Publishers

Y-Parc
Chemin de la Sallaz
CH-1400 Yverdon, Switzerland

Post Office Box 90
Reading, Berkshire RG1 8JL
Great Britain

Private Bag 8
Camberwell, Victoria 3124
Australia

3-14-9, Okubo
Shinjuku-ku, Tokyo 169
Japan

58, rue Lhomond
75005 Paris
France

Emmaplein 5
1075 AW Amsterdam
Netherlands

Glinkastrasse 13–15
O-1086 Berlin
Germany

820 Town Center Drive
Langhorne, Pennsylvania 19047
United States of America

Library of Congress Cataloging-in-Publication Data

Stochastic processes and optimal control / edited by Hans J.
Engelbert, Ioannis Karatzas, and Michael Röckner.
p. cm. -- (Stochastics monographs, ISSN 0275-5785 ; v. 7)
Lectures presented at the 9th Winterschool on Stochastic Processes and Optimal Control, held in Friedrichroda, Germany, Mar. 1-7, 1992.
Includes bibliographical references and index.
ISBN 2-88124-927-2
1. Stochastic processes--Congresses. 2. Control theory--Congresses. 3. Mathematical optimization--Congresses.
I. Engelbert, Hans Jürgen. II. Karatzas, Ioanis. III. Röckner, Michael, 1956- . IV. Winterschool on Stochastic Processes and Optimal Control (9th : 1992 : Friedrichroda, Germany) V. Series.
QA274. A1S768 1993
519.2--dc20 93-14592
CIP

 Printed in Singapore

Contents

Introduction to the Series

The journal *Stochastics and Stochastics Reports* publishes research papers dealing with stochastic processes and their applications in modelling, analysis and optimization of systems subject to random disturbances. Stochastic models are now widely used in engineering, the physical and life sciences, economics, operations research and elsewhere. Moreover, these models are becoming increasingly sophisticated and often stretch the boundaries of a theory as it exists. A primary aim of *Stochastics and Stochastics Reports* is to further the development of the field by promoting an awareness of the latest theoretical developments on the one hand and of all problems arising in applications on the other.

In association with *Stochastics and Stochastics Reports*, we are now publishing *Stochastics Monographs*, a series of independently produced volumes with the same aims and scope as the journal. *Stochastics Monographs* will provide timely and authoritative coverage of areas of current research in a more extended and expository form than is possible within the confines of a journal article. The series will include extended research reports, material derived from lecture courses on advanced topics and multi-author works with a unified theme based on conference or workshop presentations.

Mark Davis

Preface

This volume contains lectures and contributions from the 9th Winter School on Stochastic Processes and Optimal Control which was held in Friedrichroda, 1–7 March 1992. The conference was organized by the Friedrich Schiller University in Jena as part of a series of regular meetings which began in 1976 involving East German probabilists specializing in stochastic processes. The organization of this first meeting since German reunification was particularly important to continue the traditions of earlier conferences and to unite the scientific community. Therefore, it is an achievement that, despite several difficulties, the winter school could be arranged at this time.

The winter school was attended by thirty participants, among them seven foreign guests. The scientific programme included nine lectures and twenty-one seminars.

The scope of the conference was covered by aspects of the following topics:

- Dirichlet forms and stochastic processes,
- Stochastic analysis and finance mathematics,
- Stochastic differential equations,
- Stochastic partial differential equations,
- Markov processes.

The winter school included two series of lectures:

Constrained Control Problems in Finance Mathematics by I. Karatzas and M. Röckner

Dirichlet Forms and Stochastic Processes by M. Röckner.

The organizer would like to take the opportunity to thank Ioanis Karatzas and Michael Röckner for their lively and interesting lectures and their contributions to the success of the conference.

We hope that the volume will give a representative insight into the work of the winter school and that it will encourage the worldwide collaboration of probabilists working in the field of stochastic processes and their applications.

We would like to thank all speakers and participants for their contribution to both the scientific and the social success of the meeting. We also thank Karola Liese for her valuable assistance in organizing the winter school.

The winter school was supported by the Deutsche Forschungsgemeinschaft (DFG). It is a pleasure for us to express our gratitude to the DFG for this financial support without which the conference would not have taken place.

MARKOV UNIQUENESS FOR A CLASS OF INFINITE DIMENSIONAL DIRICHLET OPERATORS

SERGIO ALBEVERIO
Institut für Mathematik, Ruhr-Universität Bochum,
4630 Bochum 1, Germany

MICHAEL RÖCKNER
Institut für Angewandte Mathematik, Universität Bonn,
Wegelerstraße 6, 5300 Bonn 1, Germany

TU-SHENG ZHANG
Department of Mathematics, University of Edinburgh,
Mayfield Road, Edinburgh EH9 3JZ, Scotland, U.K.

ABSTRACT

We prove that for a class of positive finite measures μ on a locally convex topological vector space E "Markov uniqueness" holds for the corresponding Dirichlet operators $\Delta_\infty + \beta \cdot \nabla$ on $L^2(E;\mu)$. Here β is the logarithmic derivative of μ associated with a rigging $E' \subset H \subset E$ for some Hilbert space H. Markov uniqueness means that there exists exactly one self-adjoint operator L generating a sub-Markovian semigroup on $L^2(E;\mu)$ and L is on smooth cylinder functions of the above type. As a consequence we prove that we have Markov uniqueness for all positive finite measures μ on E admitting an integration by parts formula provided H is chosen properly. We also discuss applications to show uniqueness of the corresponding martingale problem under symmetry assumptions and of the stochastic quantization of infinite volume quantum fields. Furthermore, we include a complete proof for the finite dimensional case.

1990 Subject Classification
Primary: 31 C 25
Secondary: 60 J 60, 47 B 99, 81 T 08

Key Words
Generalized Schrödinger operators, uniqueness of Dirichlet operators, Dirichlet forms, (infinite dimensional) diffusions, uniqueness of the martingale problem, stochastic differential equations in infinite dimensions, Sobolev spaces on infinite dimensional spaces, stochastic quantization, quantum fields.

1 INTRODUCTION AND MAIN RESULTS

In the last part of his lectures at the Friedrichroda Winter School the second named author addressed the problem of *Markov uniqueness* and described the results that had been obtained up to that time. Meanwhile, some progress has been made. In particular, Markov uniqueness has been proved always to hold for generalized Schrödinger operators on $\mathbb{R}^d$ (cf. [RZ 92b]). The purpose of this paper is to deduce Markov uniqueness for certain generalized Schrödinger operators on infinite dimensional (Banach) spaces. Concerning the other parts of the second author's lectures in Friedrichroda, i.e., the general theory of Dirichlet forms and applications to stochastic differential equations on infinite dimensional state spaces, we refer the reader to the recent book [MR 92] resp. the article [AR 91] (also concerning more background literature). All results of this paper have been announced in [ARZ 92], at the "Conference on Measure-valued Processes, Stochastic Partial Differential Equations and Interacting Systems", Montreal, 11-16 October, 1992, and the meeting "Stochastische Analysis", Oberwolfach, 26-31 October, 1992.

Before we describe our results more precisely, we first present the finite dimensional case solved in [RZ 92a,b] (see also [R 91]) and recall our (infinite dimensional) framework.

Let dx denote Lebesgue measure on $\mathbb{R}^d$ and let φ be locally in $H^{1,2}(\mathbb{R}^d; dx)$ (= *(real) Sobolev space of order 1* in $L^2(\mathbb{R}^d; dx)$) with $\varphi \neq 0$ dx-a.e.. Consider the operator S_φ on $L^2(\mathbb{R}^d; \varphi^2 dx)$ defined by

$$S_\varphi := \Delta + 2\varphi^{-1}\langle\nabla\varphi, \nabla\cdot\rangle_{\mathbb{R}^d}$$

$$D(S_\varphi) := C_0^\infty(\mathbb{R}^d)$$

where $C_0^\infty(\mathbb{R}^d)$ denotes the set of all infinitely differentiable functions on $\mathbb{R}^d$ with compact support and Δ is the Laplacian on $\mathbb{R}^d$. Note that

$\varphi \in H^{1,2}_{loc}(\mathbb{R}^d; dx)$ is necessary and sufficient in order that $C_0^\infty(\mathbb{R}^d) \subset L^2(\mathbb{R}^d; \varphi^2 \cdot dx)$ holds and S_φ maps $C_0^\infty(\mathbb{R}^d)$ into $L^2(\mathbb{R}^d; \varphi^2 dx)$. The following was proved in [RZ 92b].

Theorem 1.1. *There exists exactly one negative definite self-adjoint operator L_φ on $L^2(\mathbb{R}^d; \varphi^2 dx)$ which extends $(S_\varphi, C_0^\infty(\mathbb{R}^d))$ such that $T_t := e^{tL_\varphi}$, $t > 0$, is sub-Markovian.*

Proof. Sect. 3 below. □

We recall that $(T_t)_{t>0}$ is called *sub-Markovian* if $0 \le u \le 1$ implies $0 \le T_t u \le 1$ for all $u \in L^2(\mathbb{R}^d; \varphi^2 \cdot dx)$, $t > 0$. L_φ *extends* $(S_\varphi, C_0^\infty(\mathbb{R}^d))$ means $C_0^\infty(\mathbb{R}^d) \subset D(L_\varphi)$ and $L_\varphi = S_\varphi$ on $C_0^\infty(\mathbb{R}^d)$. The proof of Theorem 1.1 in [RZ 92b] was based on the results in [RZ 92a]. In the proof given below we avoid that by including some modified argument extracted from [RZ 92a], [R 91]. Thus our presentation here is both complete and reasonably self-contained.

Remark 1.2. (i) The assumption $\varphi \ne 0$ dx-a.e. is not necessary and can be dropped at least if $\varphi \in H^{1,2}(\mathbb{R}^d; dx)$ (globally).

(ii) The uniqueness in Theorem 1.1 is referred to as *Markov* (or *Dirichlet) uniqueness.*

(iii) Recall that a negative definite self-adjoint operator L on $L^2(\mathbb{R}^d; \varphi^2 dx)$ generates a sub-Markovian semigroup if and only if it is a *Dirichlet operator*, that is

$$\int (u-1)^+ Lu\, \varphi^2 dx \le 0 \text{ for all } u \in D(L)\ .$$

(cf. e.g. [MR 92, Chap.I, Proposition 4.3 and Exercise 4.2(ii)]). This is in turn equivalent to the corresponding quadratic form $(\mathcal{E}, D(\mathcal{E}))$ being a *Dirichlet form* (cf. [MR 92, Chap.I, Sect. 4]). By Theorem 1.1 it follows that there exists exactly one *Dirichlet operator* L_φ on $L^2(\mathbb{R}^d; \varphi^2 \cdot dx)$ extending $(S_\varphi, C_0^\infty(\mathbb{R}^d))$ and such that the corresponding quadratic form $(\mathcal{E}_\varphi, D(\mathcal{E}_\varphi))$ is the closure on $L^2(\mathbb{R}^d; \varphi^2 \cdot dx)$ of

$$(1.1) \quad \mathcal{E}_\varphi(u,v) = \int_{\mathbb{R}^d} \langle \nabla u, \nabla v\rangle_{\mathbb{R}^d} \varphi^2 dx; \quad u, v \in C_0^\infty(\mathbb{R}^d)\ .$$

The Dirichlet form in the preceding remark can also be defined in the infinite dimensional case. Thus, it provides a tool to formulate and study the Markov uniqueness problem in infinite dimensions which we shall do now. A detailed more pedagogical presentation of the transition from the finite to the infinite dimensional case can be found in [R 90, Subsection 4a)].

Let E be a separable real Banach space with dual E' and dualization ${}_{E'}\langle\ ,\ \rangle_E$. Let $\mathcal{B}(E)$ denote its Borel σ-algebra and let μ be a finite positive measure on $(E, \mathcal{B}(E))$ with $\mathrm{supp}[\mu] = E$. We emphasize, however, that as in [RZ 92 a,b] all results below carry over to more general locally convex topological vector spaces E with possibly $\mathrm{supp}[\mu] \neq E$. But for simplicity we confine ourselves to the above special case. Define for $K \subset E'$ the linear space

$$\mathcal{F}C_b^\infty(K) := \{f(l_1, \ldots, l_m) \,|\, m \in \mathbb{N}, f \in C_b^\infty(\mathbb{R}^m), l_1, \ldots, l_m \in K\}$$

where $C_b^\infty(\mathbb{R}^m)$ denotes the set of all infinitely differentiable functions with bounded partial derivatives of all orders. Set

$$\mathcal{F}C_b^\infty := \mathcal{F}C_b^\infty(E') \,.$$

Compared with the finite dimensional case we have that E, μ, $\mathcal{F}C_b^\infty(K)$ replace $\mathbb{R}^d$, dx, $C_0^\infty(\mathbb{R}^d)$, respectively. To define the analogue of (1.1) we still need a gradient ∇. To this end fix $u = f(l_1, \ldots, l_m) \in \mathcal{F}C_b^\infty$, $z \in E$ and define for $k \in E$

$$(1.2)\ \frac{\partial u}{\partial k}(z) := \frac{d}{ds} u(z + sk)_{|s=0} = \sum_{i=1}^m \frac{\partial f}{\partial x_i}(l_1(z), \ldots, l_m(z))\ {}_{E'}\langle l_i, k\rangle_E \,.$$

Furthermore, we assume that we are given a "tangent space" H to E at each point, i.e., H is a separable real Hilbert space such that $H \subset E$ continuously and densely. Thus, identifying H with its dual H' we have

$$(1.3) \qquad E' \subset H \subset E \quad \text{continuously and densely}$$

and ${}_{E'}\langle\ ,\ \rangle_E$ coincides with the inner product $\langle\ ,\ \rangle_H$ on H when restricted to $E' \times H$. Since by (1.2) $h \mapsto \frac{\partial u}{\partial h}(z)$ is linear and continuous on H we can define $\nabla u(z)$ to be the unique element in H such that

$$\langle \nabla u(z), h\rangle_H = \frac{\partial u}{\partial h}(z) \quad \text{for all } h \in H\ (\subset E) \,.$$

Now it is possible to define a positive definite symmetric bilinear form (henceforth briefly called *form*) on (real) $L^2(E;\mu)$ by

$$\mathcal{E}^0_{\mu,H}(u,v) := \int_E \langle \nabla u, \nabla v\rangle_H d\mu; \quad u,v \in \mathcal{F}C_b^\infty \,, \tag{1.4}$$

which is densely defined, since $\mathcal{F}C_b^\infty$ is dense in $L^2(E;\mu)$. An element $k \in E$ is called *well-μ-admissible* if there exists $\beta_k \in L^2(E;\mu)$ such that

$$\int \frac{\partial u}{\partial k}\, d\mu = -\int u\beta_k \, d\mu \quad \text{for all } u \in \mathcal{F}C_b^\infty \,.$$

Assume:

(1.5) There exists a dense linear subspace K of E' ($\subset H \subset E$) consisting of well-μ-admissible elements in E.

The condition (1.5) corresponds to the assumption $\varphi \in H^{1,2}_{loc}(\mathbb{R}^d; dx)$ in the finite dimensional case (cf. [R 90, Subsection 4a)]. Then it is easy to see that the form $(\mathcal{E}^0_{\mu,H}, \mathcal{F}C_b^\infty)$ defined in (1.4) is closable on $L^2(E;\mu)$ and that its closure $(\mathcal{E}^0_{\mu,H}, D(\mathcal{E}^0_{\mu,H}))$ is a Dirichlet form on $L^2(E;\mu)$ (cf. [MR 92, Chap.II, Subsection 3b)]). We also denote the closure of ∇ with domain $D(\mathcal{E}^0_{\mu,H})$ by ∇. Let $L^0_{\mu,H}$ with domain $D(L^0_{\mu,H})$ be its generator (cf. [MR 92, Chap.I]). $(L^0_{\mu,H}, D(L^0_{\mu,H}))$ is a *Dirichlet operator*, i.e., $L^0_{\mu,H}$ is self-adjoint and $((u-1)^+, L^0_{\mu,H}u)_{L^2(E;\mu)} \leq 0$ for all $u \in D(L^0_{\mu,H})$ (cf. 1.2(iii) above and [MR 92, Chap.I, Sect. 4]) or equivalently $(e^{tL^0_{H,\mu}})_{t\geq 0}$ is sub-Markovian. It is immediate that if $u = f(l_1,\ldots,l_m) \in \mathcal{F}C_b^\infty(K)$ and $K_0 \subset K$ is an orthonormal basis of H having $l_1,\ldots,l_m$ in its linear span, then $u \in D(L^0_{\mu,H})$ and

$$L^0_{\mu,H} u = \sum_{k \subset K_0} \left(\frac{\partial}{\partial k}\left(\frac{\partial u}{\partial k}\right) + \beta_k \frac{\partial u}{\partial k}\right) . \tag{1.6}$$

Note that the right hand side is only a finite sum and defines a linear operator $S_{\mu,H,K}$ on $L^2(E;\mu)$ with dense domain $\mathcal{F}C_b^\infty(K)$. Now we can extend the notion of "Markov uniqueness" to this infinite dimensional setting.

Definition 1.3. We say that *Markov uniqueness* holds for $(S_{\mu,H,K}, \mathcal{F}C_b^\infty(K))$ if $(L^0_{\mu,H} D(L^0_{\mu,H}))$ is the only Dirichlet operator on $L^2(E;\mu)$ extending $(S_{\mu,H,K}, \mathcal{F}C_b^\infty(K))$.

For a sub-σ-algebra $\mathcal{A}$ of $\mathcal{B}$ let $E_\mu[\cdot \mid \mathcal{A}]$ denote the conditional expectation w.r.t. μ given $\mathcal{A}$. Now we can formulate the main result of this paper.

Theorem 1.4. *Suppose that there exist $k_i \in K$, $i \in \mathbb{N}$, forming an orthonormal basis in H such that for $\mathcal{B}_N := \sigma\{_{E'}\langle k_i, \cdot \rangle_E \mid i \leq N\}$*

$$\sup_{N\in\mathbb{N}} \sum_{i=1}^{N} \|\beta_{k_i} - E_\mu[\beta_{k_i} \mid \mathcal{B}_N]\|^2_{L^2(E;\mu)} < \infty \; . \tag{1.7}$$

Then Markov uniqueness holds for $(S_{\mu,H,K}, \mathcal{F}C_b^\infty(K))$.

Proof. Sect. 4 below. □

The condition (1.7) is a weakened form of a condition in [T 87] (see also [AH-K 77] and [T 85]) where an analogue of Theorem 1.4 is proved under much stronger smoothness assumptions on μ (cf. [T 87, Theorem 3]). The proof of Theorem 1.4 is based on Theorem 1.1 above and the following regularity result which in turn is a consequence of the proof of the "Fukushima conjecture" in [AR 90, cf. Theorem 5.2], or [Ma 78, Part I, §1].

Proposition 1.5. *Let ν be a finite positive measure on $(\mathbb{R}^d, \mathcal{B}(\mathbb{R}^d))$. Assume that for every $1 \leq i \leq d$ there exists $\beta_i \in L^2(\mathbb{R}^d; \nu)$ such that*

$$\int \frac{\partial u}{\partial x_i}\, d\nu = -\int u\beta_i \, d\nu \quad \textit{for all } u \in C_0^\infty(\mathbb{R}^d) \; .$$

Then there exists $\varphi \in H^{1,2}(\mathbb{R}^d; dx)$ such that $\nu = \varphi^2 \cdot dx$.

Proof. Sect. 3 below. □

As easy consequences of Theorem 1.4 we obtain

Corollary 1.6. *Consider the situation of Theorem 1.4; i.e., in particular (1.7) holds. Let $\varphi \in D(\mathcal{E}^0_{\mu,H})$ such that $\varphi\beta_k \in L^2(E;\mu)$ for all $k \in K$. Then Markov uniqueness holds for $(S_{\mu,H,K} + 2\varphi^{-1}\langle\nabla\varphi, \nabla\cdot\rangle_H$, $\mathcal{F}C_b^\infty(K))$ (on $L^2(E;\varphi^2\mu)$).*

Proof. Sect. 4 below. □

Corollary 1.7. *There exists a separable real Hilbert space $\tilde{H} \subset H$ and a linear subspace $\tilde{K} \subset K$ such that the image of $\tilde{K}$ under the imbedding $\tilde{K} \subset E' \subset \tilde{H}' \equiv \tilde{H} \subset E$ consists of well-μ-admissible elements in E and Markov uniqueness holds for the corresponding operator $(S_{\mu,\tilde{H},\tilde{K}}, \mathcal{F}C_b^\infty(\tilde{K}))$.*

Proof. See Sect. 4 below. □

Now we turn to the probabilistic consequences of 1.4, 1.7. By [Sch 90] (or [MR 92, Chap.IV, Subsection 4b)]) there exists a diffusion process $\mathbf{M}_{\mu,H} = (\Omega, \mathcal{T}, (\mathcal{T}_t), (X_t), (P_z)_{z\in E})$ *associated with* the Dirichlet form $(\mathcal{E}^0_{\mu,H}, D(\mathcal{E}^0_{\mu,H}))$, i.e., for all $u \in L^2(E;\mu)$, $t > 0$,

$$\int u(X_t)dP_z = T_t u(z) \quad \text{for } \mu\text{-a.e. } z \in E$$

where $T_t u := \exp(tL^0_{H,\mu})u$. $\mathbf{M}_{\mu,H}$ is easily seen to be conservative (i.e., has infinite lifetime) and to have μ as an invariant measure. For the definition of *diffusion processes* and *right processes* below we refer to [MR 92, Chap.IV, Sect.1] and also [Sh 88]. A right process is called *μ-symmetric* if its transition group is μ-symmetric.

Definition 1.8. Let $D \subset L^2(E;\mu)$ such that each $u \in D$ has a μ-version $\tilde{u}$ which is continuous. Let A be a linear operator on $L^2(E;\mu)$ having D in its domain. We say that a μ-symmetric right process $\mathbf{M} = (\Omega, \mathcal{F}, (\mathcal{F}_t), (X_t), (P_z)_{z\in E})$ with state space E solves the martingale problem for (A, D) if for every $u \in D$

$$\tilde{u}(X_t) - \tilde{u}(X_0) - \int_0^t Au(X_s)ds,\ t \geq 0, \tag{1.8}$$

is an $(\mathcal{F}_t)_{t\geq 0}$-martingale under $P_\mu := \int P_z \mu(dz)$.

Note that 1.8 is a modification of [RZ 92a, Definition 6.1]. Note also that $\int_0^t Au(X_s)ds$ is independent of which μ-version of Au we take, since for any $\mathcal{B}(E)$-measurable, μ-square integrable function f we have by μ-symmetry that

$$\begin{aligned}\int\int_0^t |f|(X_s)ds\, dP_\mu &\leq e^t \int_0^\infty e^{-s} \int 1_E(X_s)|f|(X_0)dP_\mu ds \\ &\leq e^t \int |f| d\mu\end{aligned}$$

where 1_E is the indicator function of E. We emphasize that the right process in 1.8 might have a lifetime $\zeta \not\equiv \infty$. As usual an isolated point Δ is adjoined to E such that $X_t = \Delta$ if $t \geq \zeta$ and any function $f : E \to \mathbb{R} \cup \{\pm\infty\}$ is extended to E_Δ by $f(\Delta) = 0$, so that (1.8) makes sense for all $t \geq 0$ (cf. [MR 92, Chap.IV, Sect.1]). By [MR 92, Chap.IV, Theorem 6.4] the following is then obvious, but nevertheless for the convenience of the reader we include a proof in Sect. 4 below.

Theorem 1.9. *Suppose Markov uniqueness holds for* $(S_{\mu,H,K}, \mathcal{F}C_b^\infty(K))$. *Then there exists (up to* μ*-equivalence) exactly one* μ*-symmetric right process* $\mathbf{M}$ *with state space* E *which solves the martingale problem for* $(S_{\mu,H,K}, \mathcal{F}C_b^\infty(K))$. $\mathbf{M}$ *is, in fact, the conservative diffusion process* $\mathbf{M}_{\mu,H}$ *introduced above. This holds, in particular, in the situations described in Theorem 1.4 and Corollaries 1.6, 1.7.*

We conclude this section with the following remark and examples.

Remark 1.10. In the situation described before Theorem 1.4 it was proved in [AR 91, cf. Theorem 6.6] that if $\int {}_{E'}\langle k, z\rangle_E^2 \mu(dz) < \infty$ for all $k \in K$ and if there exists a Brownian semigroup on E over H (i.e. with covariance given by $2\langle\ ,\ \rangle_H$) then $\mathbf{M}_{\mu,H}$ satisfies the stochastic equation

$$X_t = z + W_t + N_t, \quad t \geq 0, P_z\text{-a.s.} \tag{1.9}$$

for $\mathcal{E}^0_{\mu,H}$-quasi-every $z \in E$ (cf. [MR 92, Chap.III, Sect.2] for the notion of $\mathcal{E}^0_{\mu,H}$-quasi-everywhere). Here $(W_t)_{t\geq 0}$ is an $(\mathcal{F}_t)$-Brownian motion on E with covariance $2\langle\ ,\ \rangle_H$ starting at the origin under P_z and $(N_t)_{t\geq 0}$ is a continuous E-valued $(\mathcal{F}_t)$-adapted process such that for every $k \in K$, ${}_{E'}\langle k, N_t\rangle_E = \int_0^t \beta_k(X_s)ds$, $t \geq 0$, P_z-a.s. for $\mathcal{E}^0_{\mu,H}$-quasi-every $z \in E$. By Itô's formula (in finite dimensions!) it follows from Theorem 1.9 that $\mathbf{M}_{\mu,H}$ is (up to μ-equivalence) the only μ-symmetric right process satisfying (1.9) if Markov uniqueness holds for $(S_{\mu,H,K}, \mathcal{F}C_b^\infty(K))$.

Example 1.11. (i) Corollary 1.7 obviously applies, in particular, to the examples studied in detail in [AR 91, Subsection 7 II]. There μ was an (*infinite volume*) *space time* resp. *time zero quantum field* with E a Banach space of tempered Schwartz distributions on $\mathbb{R}^d$, $H := L^2(\mathbb{R}^d; dx)$ and $K := C_0^\infty(\mathbb{R}^d)$ with $d = 2$ resp. $d = 1$. We emphasize, however, that it seems unlikely that (1.7) holds for these E, H, μ. Therefore, to have uniqueness we change the Hilbert space H into $\tilde{H}$, and K into $\tilde{K}$ as in 1.7.. Nevertheless, we have thus constructed a solution $\mathbf{M}_{\mu,\tilde{H}}$ of (1.9) (with $\tilde{H}, \tilde{K}$) which is within the class of μ-symmetric right processes (up to μ-equivalence) unique, is conservative, and has μ as an invariant measure. $\mathbf{M}_{\mu,\tilde{H}}$ is sometimes called the *stochastic quantization* of μ.

(ii) A special example of Corollary 1.6 was proved in [RZ 92b, Sect. 4], namely the case where μ is a Gaussian measure and H is not necessarily the reproducing kernel Hilbert space of μ, but such that the covariance operator "considered on H" has discrete spectrum.

2 PRELIMINARIES

Consider the situation described in Sect.1. In particular, assumption (1.5) is still in force for some $K \subset E'$ ($\subset H' \equiv H \subset E$). For $k \in E\backslash\{0\}$, choose a closed subspace E_k of E such that $E = E_k \oplus k\mathbb{R}$ and let $\pi_k : E \to E_k$ be the canonical projection. Then there exists a kernel

$$\overline{\rho}_k : E \times \mathcal{B}(\mathbb{R}) \to [0,1]$$

such that

$$L^2(E;\mu) = \int^{\oplus} L^2(\mathbb{R};\overline{\rho}_k(x,ds))\nu_k(dx), \tag{2.1}$$

in the sense that $u \in L^2(E;\mu)$ corresponds to a *measurable field of vectors* (w.r.t. μ) $(u_x)_{x\in E_k}$ where $u_x := u(x+\cdot k)$, $x \in E_k$ (cf. [AR 90]). Here $\nu_k := \mu \circ \pi_k^{-1}$, i.e., the image measure of μ under π_k.
For a function $\rho : \mathbb{R} \to [0,+\infty[$, the *regular set* $R(\rho)$ of ρ is defined by

$$R(\rho) := \{s \in \mathbb{R} \mid \int_{s-\varepsilon}^{s+\varepsilon} \frac{1}{\rho(s)} ds < +\infty \text{ for some } \varepsilon > 0\} .$$

If $\rho = 0$ ds-a.e. on $\mathbb{R}\backslash R(\rho)$, we say ρ satisfies *Hamza's condition.*
Let $k \in E\backslash\{0\}$ be well-μ-admissible. We have that for ν_k-a.e. $x \in E_k$, $\overline{\rho}_k(x,ds) = \rho_k(x,s)ds$. The function $s \mapsto \rho_k(x,s)$ can be assumed to be absolutely continuous, it satisfies Hamza's condition, and $\beta_k = (\frac{\partial}{\partial s}\rho_k(x,\cdot)/\ \rho_k(x,\cdot))_{x\in E_k}$ (cf. [RZ 92a, Sect.1] for more details). By [AKR 90, Proposition 2.2] and [AR 90, Theorem 3.2] we know that the following form $(\mathcal{E}^+_{\mu,k}, D(\mathcal{E}^+_{\mu,k}))$ on $L^2(E;\mu)$ is closed and is, moreover, a (symmetric) Dirichlet form.

$$\begin{aligned} D(\mathcal{E}^+_{\mu,k}) := \{u = (u_x)_{x\in E_k} \in \int^{\oplus} & L^2(\mathbb{R};\rho_k(x,s)ds)\nu_k(dx) \mid \\ & \text{for } \nu_k \text{ a.e. } x \in E_k,\ u_x \text{ has an absolutely continuous} \\ & (ds)\text{-version } \tilde{u}_x \text{ on } R(\rho_k(x,\cdot)) \text{ such that} \\ & \frac{\partial u}{\partial k} := (\frac{d\tilde{u}_x}{ds})_{x\in E_k} \in \int^{\oplus} L^2(\mathbb{R};\rho_k(x,s)ds)\nu_k(dx)\}, \end{aligned}$$

$$\mathcal{E}^+_{\mu,k}(u,v) := \int \frac{\partial u}{\partial k}\frac{\partial v}{\partial k} d\mu;\ u,v \in D(\mathcal{E}^+_{\mu,k}) .$$

Remark 2.1. Clearly, if $u, v \in D(\mathcal{E}^+_{\mu,k})$ such that $uv, v\frac{\partial u}{\partial k}, u\frac{\partial v}{\partial k} \in L^2(E;\mu)$, then $uv \in D(\mathcal{E}^+_{\mu,k})$ and $\frac{\partial}{\partial k}(uv) = u\frac{\partial v}{\partial k} + v\frac{\partial u}{\partial k}$.

Lemma 2.2. *Let $k \in E\backslash\{0\}$ be well-μ-admissible and $u \in L^2(E;\mu)$. Then $u \in D(\mathcal{E}^+_{\mu,k})$ if and only if there exists $f \in L^2(E;\mu)$ such that*

$$\int u\frac{\partial v}{\partial k}\,d\mu = -\int (f + u\beta_k)v\,d\mu \quad \textit{for all } v \in \mathcal{F}C_b^\infty\ . \tag{2.2}$$

In this case $f = \frac{\partial u}{\partial k}$.

Proof. (Cf. [RZ 92a, Remark 1.5(i)], but note that there is a misprint in the formula of $\frac{\partial u}{\partial k}$ there). By [AKR 90, Theorem 2.5] we know that if $u \in D(\mathcal{E}^+_{\mu,k})$, then

$$\int \frac{\partial u}{\partial k} v\,d\mu = -\int u(\frac{\partial v}{\partial k} + v\beta_k)d\mu \quad \text{for all } v \in \mathcal{F}C_b^\infty\ .$$

Conversely, if u satisfies (2.2) then by disintegration (i.e., (2.1)) it follows that no matter how we choose the complementing subspace E_k there exists $\Omega \in \mathcal{B}(E_k)$ with $\nu_k(\Omega) = 1$ such that for all $x \in \Omega$, $g \in C_0^\infty(\mathbb{R})$

$$\begin{aligned} &-\int (f(x+sk) + u(x+sk)\beta_k(x+sk))g(s)\rho_k(x,s)ds \\ &= \int u(x+sk)\frac{dg}{ds}\rho_k(x,s)ds\ . \end{aligned} \tag{2.3}$$

By [Mi 73, Theorem 2.7] we conclude that for ν_k-a.e. $x \in E_k$, $s \mapsto u(x+sk)\rho_k(x,s)$ has an absolutely continuous ds-version on $\mathbb{R}$. By [AKR 90, Proposition 2.4(i)] $R(\rho_k(x,\cdot)) = \{\rho_k(x,\cdot) > 0\}$ (recall $\rho_k(x,\cdot)$ is absolutely continuous), hence $s \mapsto u(x+sk)$ has an absolutely continuous ds-version $\tilde{u}_x$ on $R(\rho_k(x,\cdot))$ for ν_k-a.e. $x \in E_k$. We now see by (2.3) that

$$(\frac{\partial \tilde{u}_x}{ds})_{x\in E_k} = (f(x+\cdot k))_{x\in E_k},$$

i.e., $\frac{\partial u}{\partial k} = f$. □

Let us recall that there is an order $\prec$ on forms on $L^2(E;\mu)$ defined by

$$(\mathcal{E}^1, D(\mathcal{E}^1)) \prec (\mathcal{E}^2, D(\mathcal{E}^2))$$

if

$$D(\mathcal{E}^1) \subset D(\mathcal{E}^2) \text{ and } \mathcal{E}^1(u,u) \geq \mathcal{E}^2(u,u) \text{ for } u \in D(\mathcal{E}^1)\ .$$

In addition, to the form $(\mathcal{E}^0_{\mu,H}, D(\mathcal{E}^0_{\mu,H}))$ introduced in Sect.1 there is another natural form on $L^2(E;\mu)$ in our situation defined by

$$D(\mathcal{E}^+_{\mu,H,K}) := \{u \in \bigcap_{k\in K} D(\mathcal{E}^+_{\mu,k}) \,|\, \text{there exists a } \mathcal{B}(E)/\mathcal{B}(H)$$
$$\text{- measurable map } \nabla u : E \to H \text{ such that for each}$$
$$k \in K,\ \langle \nabla u(z), k\rangle_H = \frac{\partial u}{\partial k}(z) \text{ for } \mu\text{-a.e. } z \in E \text{ and}$$
$$\int \langle \nabla u, \nabla u\rangle_H d\mu < +\infty\}$$
$$\mathcal{E}^+_{\mu,H,K}(u,v) := \int_E \langle \nabla u, \nabla v\rangle_H d\mu;\quad u,v \in D(\mathcal{E}^+_{\mu,H,K}).$$

By [AR 90, Theorem 3.10] $(\mathcal{E}^+_{\mu,H,K}, D(\mathcal{E}^+_{\mu,H,K}))$ is a Dirichlet form which extends the closure of $(\mathcal{E}^0_{\mu,H}, D(\mathcal{E}^0_{\mu,H}))$. We denote its generator by $L^+_{\mu,H,K}$. Now we define for $S_{\mu,H,K}$ as in (1.6)

$$\mathcal{A}_D(S_{\mu,H,K}) := \{L \,|\, L \text{ is a Dirichlet operator on } L^2(E;\mu) \text{ which extends } (S_{\mu,H,K}, \mathcal{F}C_b^\infty(K))\}.$$

For $L_1, L_2 \in \mathcal{A}_D(S_{\mu,H,K})$, we define $L_1 \prec L_2$ if for the corresponding Dirichlet forms we have $(\mathcal{E}_1, D(\mathcal{E}_1)) \prec (\mathcal{E}_2, D(\mathcal{E}_2))$. The following theorem will be used (see [AKR 90, Theorem 3.1]).

Theorem 2.3. *$L^+_{\mu,H,K}$ is the maximal element in $\mathcal{A}_D(S_{\mu,H,K})$ with respect to $\prec$ defined above and $L^0_{\mu,H}$ is the minimal one in $\mathcal{A}_D(S_{\mu,H,K})$. In particular, Markov uniqueness holds for $(S_{\mu,H,K}, \mathcal{F}C_b^\infty(K))$ if and only if $D(\mathcal{E}^0_{\mu,H}) = D(\mathcal{E}^+_{\mu,H,K})$.*

Remark 2.4. Because of 2.3, analogously to the case $E = \mathbb{R}^d$, $\mu = dx$, in [RZ 92a] the spaces $D(\mathcal{E}^0_{\mu,H})$, $D(\mathcal{E}^+_{\mu,H,K})$ were denoted by $H_0^{1,2}(E;\mu)$ resp. $H^{1,2}(E;K;\mu)$, i.e., the minimal resp. maximal Sobolev spaces of order 1 over $L^2(E;\mu)$.

3 PROOFS FOR THE RESULTS IN FINITE DIMENSIONS

In this section we shall prove Theorem 1.1 (including Remark 1.2(i)) and also Proposition 1.5. Hence we consider the case $E = H = \mathbb{R}^d = K$ and fix $\varphi \in H^{1,2}_{loc}(\mathbb{R}^d; dx))$, i.e., $\varphi \in L^2_{loc}(\mathbb{R}^d; dx)$, $\nabla\varphi \in L^2_{loc}(\mathbb{R}^d \to \mathbb{R}^d; dx)$. (Here $\nabla\varphi$ is defined in the sense of Schwartz distributions).

Set $\mu := \varphi^2 \cdot dx$. We adopt the terminology of Section 2 though μ is not finite, but we replace $\mathcal{F}C_b^\infty(\mathbb{R}^d)$ by $C_0^\infty(\mathbb{R}^d)$. Denote $(\mathcal{E}^0_{\mu,H}, D(\mathcal{E}^0_{\mu,H}))$ by $(\mathcal{E}^0_\varphi, D(\mathcal{E}^0_\varphi))$, i.e., $(\mathcal{E}^0_\varphi, D(\mathcal{E}^0_\varphi))$ is the closure on $L^2(\mathbb{R}^d; \varphi^2 \cdot dx)$ of

$$\mathcal{E}^0_\varphi(u,v) = \int_{\mathbb{R}^d} \langle \nabla u, \nabla v\rangle_{\mathbb{R}^d}\ \varphi^2 dx; \quad u,v \in C_0^\infty(\mathbb{R}^d)\ . \tag{3.1}$$

We also set $(\mathcal{E}^+_\varphi, D(\mathcal{E}^+_\varphi)) := (\mathcal{E}^+_{\mu,H,K}, D(\mathcal{E}^+_{\mu,H,K}))$ and we may assume that $\varphi \geq 0$. The reason, why μ is assumed to be absolutely continuous w.r.t. dx a priori, is that in order to have closability of the components of (3.1) (with μ replacing $\varphi^2 \cdot dx$) on $L^2(\mathbb{R}^d; \mu)$ by [AR 90, Theorem 5.2] we must have that $\mu \ll dx$. For the proof of Theorem 1.1 we need to recall the notion of *$\mathcal{E}^0_\varphi$-nest* and *$\mathcal{E}^0_\varphi$-quasi-continuity* (cf. [MR 92, Chap.III, Sections 2, 3]). An increasing sequence $(E_n)_{n\in\mathbb{N}}$ of closed sets in $\mathbb{R}^d$ is called an *$\mathcal{E}^0_\varphi$-nest* if

$$\bigcup_{n\in\mathbb{N}} \{u \in D(\mathcal{E}^0_\varphi)\,|\, u = 0 \text{ on } E_n^c \quad \varphi^2 \cdot dx\text{-a.e.}\}$$

is dense in $D(\mathcal{E}^0_\varphi)$ w.r.t. $(\mathcal{E}^0_{\varphi,1})^{1/2}$ where $\mathcal{E}^0_{\varphi,1} := \mathcal{E}^0_\varphi + (\ ,\ )_{L^2(\mathbb{R}^d;\varphi^2\cdot dx)}$ and for $A \subset \mathbb{R}^d$, $A^c := \mathbb{R}^d \backslash A$. A function $u : \mathbb{R}^d \to \mathbb{R}$ is called *$\mathcal{E}^0_\varphi$-quasi-continuous* if there exists an $\mathcal{E}^0_\varphi$-nest $(E_n)_{n\in\mathbb{N}}$ such that $u_{|E_n}$ (= restriction of u to E_n) is continuous for all $n \in \mathbb{N}$. Recall that by [MR 92, Chap.IV, Subsection 4a)] there always exists an $\mathcal{E}^0_\varphi$-nest of compacts since $(\mathcal{E}^0_\varphi, D(\mathcal{E}^0_\varphi))$ is *quasi-regular* (cf. [MR 92, Chap.IV, Def.3.1]). Hence, by making the closed sets smaller if necessary every $\mathcal{E}^0_\varphi$-nest can be assumed to consist of compacts. Below we often quote [MR 92] since it is a little bit more convenient for our definition of $\mathcal{E}^0_\varphi$-nest above. Corresponding facts can, of course, also be found in [F 80]. Let $(\mathcal{E}, H^{1,2}(\mathbb{R}^d))$ be the Dirichlet form associated with d-dimensional Brownian motion, i.e., $(\mathcal{E}, H^{1,2}(\mathbb{R}^d)) = (\mathcal{E}^0_\varphi, D(\mathcal{E}^0_\varphi))$ for $\varphi \equiv 1$.

Proposition 3.1. *Every $\mathcal{E}$-nest $(K_n)_{n\in\mathbb{N}}$ is also an $\mathcal{E}^0_\varphi$-nest.*

Proof. Let $\mathbf{M}^\varphi := (\Omega, \mathcal{F}, (\mathcal{F}_t), (X_t), (P_x)_{x\in\mathbb{R}^d})$ be the $(\varphi^2 \cdot dx)$-symmetric conservative diffusion process associated with the Dirichlet form $(\mathcal{E}^0_\varphi, D(\mathcal{E}^0_\varphi))$ (cf. [F 80, Sect. 6] or [MR 92, Chap.IV]). We may asume that $\Omega = C([0,\infty[, \mathbb{R}^d)$, $X_t : \Omega \to \mathbb{R}^d$ is evaluation at $t \geq 0$ and $(\mathcal{F}_t)$ is the *minimum completed admissible family* (cf. [F 80, p.89]). By the Fukushima decomposition (see [F 80, Theorem 5.2.2] and [AR 91, Sect.5]) we have

$$X_t = x + B_t + \int_0^t 2\frac{\nabla\varphi}{\varphi}(X_s)ds,\ t \geq 0,\ P_x\text{-a.e. for } dx\text{-a.e. } x \in \mathbb{R}^d\ ,$$

where (B_t) is an $(\mathcal{F}_t)$-Brownian motion on $\mathbb{R}^d$ starting at the origin. Note that $\|\frac{\nabla\varphi}{\varphi}\|_{\mathbb{R}^d} \in L^2(\mathbb{R}^d;\varphi^2 dx)$. Hence, if $(R_\alpha^\varphi)_{\alpha>0}$ is the resolvent of $\mathbf{M}^\varphi$ we have by $(\varphi^2\cdot dx)$-symmetry

$$\begin{aligned}\int P_x[\int_0^t \frac{\|\nabla\varphi\|_{\mathbb{R}^d}^2}{\varphi^2}(X_s)ds]\varphi^2 dx &\leq e^t \int R_1^\varphi(\frac{\|\nabla\varphi\|_{\mathbb{R}^d}^2}{\varphi^2})\varphi^2 dx \\ &= e^t \int R_1^\varphi 1 \ \ \|\nabla\varphi\|_{\mathbb{R}^d}^2 dx \\ &= e^t \int \|\nabla\varphi\|_{\mathbb{R}^d}^2 dx < \infty .\end{aligned}$$

Hence

$$\int_0^t \frac{\|\nabla\varphi\|_{\mathbb{R}^d}^2}{\varphi^2}(X_s)ds \ < \ +\infty,\ t\geq 0,\ P_x\text{-a.e. for } dx\text{-a.e. } x\in\mathbb{R}^d .$$

By [KZ 71] this implies that for dx-a.e. $x\in\mathbb{R}^d$, $P_x \ll Q_x$ on $\mathcal{F}_t^0 := \sigma(X_s\,|\,s\leq t)$ for all $t\geq 0$ where Q_x denotes the distribution of Brownian motion starting at x. The result in [KZ 71] was stated in the case $d=1$, but it obviously remains true for any finite d by using Girsanov's theorem for $\mathbb{R}^d$; (cf. e.g. [IW 89, p.192/193] or [LS 77, Chap.7]). Hence

$$P_{\varphi^2\cdot dx} \ll Q_{dx} \ \ (:= \int Q_x dx) \text{ on } \mathcal{F}_t^0 \text{ for all } t\geq 0. \tag{3.2}$$

By [MR 92, Chap.IV, Proposition 5.30] we know that for all $t\geq 0$

$$Q_{dx}[\lim_{n\to\infty}\sigma_{K_n^c} < t] \leq Q_{dx}[\lim_{n\to\infty}\sigma_{K_n^c} < \infty] = 0,$$

where for $A\subset\mathbb{R}^d$, $\sigma_A := \inf\{t>0\,|\,X_t\in A\}$. Since each $\sigma_{K_n^c}$ is an $(\bigcap_{s>t}\mathcal{F}_s^0)$-stopping time, (3.2) implies that

$$P_{\varphi^2\cdot dx}[\lim_{n\to\infty}\sigma_{K_n^c} < \infty] = 0 .$$

Hence, by [MR 92, p.128 and Chap.III, Theorem 2.11(i)] $(K_n)_{n\in\mathbb{N}}$ is an $\mathcal{E}_\varphi^0$-nest. □

Proof of Theorem 1.1, Remark 1.2(i) .
In our present situation (1.5) is equivalent with $\varphi\in H^{1,2}(\mathbb{R}^d;dx)$ and Theorem 2.3 applies in this case. Note also that then for the canonical basis $k_1,\ldots,k_d$ of $\mathbb{R}^d$

$$\beta_{k_i} = 2\varphi^{-1}\frac{\partial\varphi}{\partial x_i},\ \ 1\leq i\leq d, \tag{3.3}$$

where $\frac{\partial}{\partial x_i}$ means *ordinary* weak partial derivative. Theorem 2.3 has, however, been generalized by M. Takeda [T 92] to the case where $\varphi \in H^{1,2}_{loc}(\mathbb{R}^d; dx)$ if $\varphi \neq 0$ dx-a.e.. Hence both Theorem 1.1 and Remark 1.2(i) are proved if we can show that

$$C_0^\infty(\mathbb{R}^d) \text{ is dense in } D(\mathcal{E}_\varphi^+) \text{w.r.t. } \mathcal{E}_{\varphi,1}^+ (= \mathcal{E}_\varphi^+ + (\ ,\ )_{L^2(\mathbb{R}^d;\varphi^2 dx)}).$$

(Note that in all what follows we do not assume $\varphi \neq 0$ dx-a.e.) By a simple localization argument (cf. the last part of the proof of [RZ 92a, Theorem 4.1] with $p = q = 2$), we can assume $\varphi \in H^{1,2}(\mathbb{R}^d)$ (i.e., $\varphi \in L^2(\mathbb{R}^d; dx)$, $\nabla\varphi \in L^2(\mathbb{R}^d \to \mathbb{R}^d; dx)$). In particular, we then have that $1 \in D(\mathcal{E}_\varphi^0)$. Now, fix $u \in D(\mathcal{E}_\varphi^+)$. Since $(\mathcal{E}_\varphi^+, D(\mathcal{E}_\varphi^+))$ is a Dirichlet form, we may assume that u is bounded. The proof proceeds in several steps. Let $b_l \in C_0^\infty(\mathbb{R})$, $l \in \mathbb{N}$, with $|b_l'| \leq 2$ and $1_{[-l,l]} \leq b_l \leq 1_{]-l-1,l+1[}$. Define

$$\varphi_l := b_l(\ln \varphi),\ \ l \in \mathbb{N}.$$

Step 1. $\varphi_l \in D(\mathcal{E}_\varphi^+)$, $l \in \mathbb{N}$, and $\frac{\partial \varphi_l}{\partial k_i} = \frac{\partial \varphi_l}{\partial x_i} = b_l'(ln\varphi)\frac{\partial \varphi}{\partial x_i}\varphi^{-1}$ $\varphi^2 dx$-a.e., $1 \leq i \leq d$. Furthermore, $u\varphi_l \in D(\mathcal{E}_\varphi^+)$ and $\mathcal{E}_\varphi^+(u\varphi_l, u\varphi_l) \leq 2\mathcal{E}_\varphi^+(u,u) + 2\|u\|_\infty^2 \mathcal{E}(\varphi,\varphi)$.

Fix $l \in \mathbb{N}$, $1 \leq i \leq d$. Note first that (by definition of b_l) $\varphi_l \in H^{1,2}(\mathbb{R}^d; dx)$. Hence by the product rule for $\frac{\partial}{\partial x_i}$ and integration by parts w.r.t. dx we have for all $v \in C_0^\infty(\mathbb{R}^d)$

$$\begin{aligned}\int \frac{\partial v}{\partial x_i}\varphi_l \varphi^2 dx &= \int \frac{\partial (v\varphi_l)}{\partial x_i}\varphi^2 dx - \int v \frac{\partial \varphi_l}{\partial x_i}\varphi^2 dx \\ &= -\int v(\varphi_l \beta_{k_i} + \frac{\partial \varphi_l}{\partial x_i})\varphi^2 dx\ .\end{aligned}$$

Hence Step 1 follows by Lemma 2.2.

Step 2. $u\varphi_l \in H^{1,2}(\mathbb{R}^d; dx)$, $l \in \mathbb{N}$.

Let $l \in \mathbb{N}$, $1 \leq i \leq d$. In this part of the proof we denote the Lebesgue measure on $\mathbb{R}^m$ by λ^m. Note that in the present situation

$$\nu_{k_i} = f \cdot \lambda^{d-1}\ ,$$

where ν_{k_i} is as in (2.1) and

$$f(x_1, \ldots, x_{i-1}, x_{i+1}, \ldots, x_d) = \int \varphi^2(x_1, \ldots, x_d) dx_i\ .$$

Given $g : \mathbb{R}^d \to \mathbb{R}$ and $x = (x_1, \ldots, x_{d-1}) \in \mathbb{R}^{d-1}$ define

$$g_x^{(i)}(s) := g(x_1, \ldots, x_{i-1}, s, x_i, \ldots, x_{d-1}), \ s \in \mathbb{R}\,.$$

Since $\varphi \in H^{1,2}(\mathbb{R}^d; \lambda^d)$ there exists a λ^d-version, again denoted by φ, such that $\varphi_x^{(i)}$ is absolutely continuous for every $x \in \mathbb{R}^{d-1}$ (cf. [Mi 73, Chap. 2]). Since $u \in D(\mathcal{E}_\varphi^+)$, for ν_k-a.e. $x \in \mathbb{R}^{d-1}$ there exists a λ^1-version $\tilde{u}_x^{(i)}$ of $u_x^{(i)}$ which is absolutely continuous on $R((\varphi_x^{(i)})^2) = \{\varphi_x^{(i)} \neq 0\}$ (cf. [AKR 90, Proposition 2.4(i)]). It is obvious that for λ^{d-1}-a.e. $x \in \{f > 0\}$,

$$\psi_x := \tilde{u}_x^{(i)} b_l(ln\varphi_x^{(i)})$$

is absolutely continuous on $\mathbb{R}$. But for $x \in \{f = 0\}$, $\varphi_x^{(i)}(s) = 0$ for all $s \in \mathbb{R}$. Hence ψ_x is absolutely continuous for λ^{d-1}-a.e. $x \in \mathbb{R}^{d-1}$. Consequently, for all $v_1 \in C_0^\infty(\mathbb{R})$, $v_2 \in C_0^\infty(\mathbb{R}^{d-1})$ and $v := v_1 \otimes v_2$

$$\int \frac{\partial}{\partial k_i}(v u \varphi_l) d\lambda^d = \int v_2 \int \frac{d}{ds}(v_1 \psi_x) ds \lambda^{d-1}(dx) = 0$$

and by the product rule for $\frac{\partial}{\partial k_i}$ (cf. Remark 2.1) we thus obtain, since $u\varphi_l \in D(\mathcal{E}_\varphi^+)$ by Step 1,

$$(3.4) \int \frac{\partial v}{\partial x_i} u \varphi_l \, d\lambda^d = \int_{\{\varphi \neq 0\}} \frac{\partial}{\partial k_i}(v u \varphi_l) \, d\lambda^d - \int_{\{\varphi \neq 0\}} v \frac{\partial (u\varphi_l)}{\partial k_i} \, d\lambda^d\,.$$

But again by Step 1

$$\frac{\partial}{\partial k_i}(v u \varphi_l) = \varphi_l \frac{\partial}{\partial k_i}(vu) + v u b_l'(\ln\varphi) \frac{\partial \varphi}{\partial x_i} \varphi^{-1} = 0 \quad \text{on } \{\varphi = 0\},$$

which implies that the corresponding integral in (3.4) is equal to zero. Furthermore, by 2.1

$$1_{\{\varphi \neq 0\}} \frac{\partial}{\partial k_i}(u\varphi_l) = 1_{\{\varphi \neq 0\}}(u b_l'(\ln\varphi) \frac{\partial \varphi}{\partial x_i} \varphi^{-1} + b_l(\ln\varphi) \frac{\partial u}{\partial k_i}) \in L^2(\mathbb{R}^d; \lambda^d).$$

Hence Step 2 follows.

Step 3. $u \in D(\mathcal{E}_\varphi^0)$.

Since clearly $u\varphi_l \longrightarrow_{l\to\infty} u$ in $L^2(\mathbb{R}^d; \varphi^2 dx)$ and $\sup_{l\in\mathbb{N}} \mathcal{E}_\varphi^+(u\varphi_l, u\varphi_l) < \infty$ by Step 1, it follows by the Banach-Saks theorem (cf. e.g. [MR 92, Chap.I, Lemma 2.12]) that the Cesaro mean of a subsequence of

$(u\varphi_l)_{l\in\mathbb{N}}$ converges to u w.r.t. $\mathcal{E}^+_{\varphi,1}$. Hence by Step 2 we may assume that $u \in H^{1,2}(\mathbb{R}^d; dx)$. Let $\tilde{\varphi}$, $\tilde{u}$ denote $\mathcal{E}$-quasi-continuous dx-versions of φ, u respectively. Let $(K_n)_{n\in\mathbb{N}}$ be an $\mathcal{E}$-nest of compacts such that $\tilde{\varphi}_{|K_n}$ and $\tilde{u}_{|K_n}$ are continuous for all $n \in \mathbb{N}$. For $n \in \mathbb{N}$ let e_n denote the *1-equilibrium potential* of the set K_n^c w.r.t. $(\mathcal{E}^0_\varphi, D(\mathcal{E}^0_\varphi))$ (cf. [F 80, §3.1] or [MR 92, Chap.III, Sect.1]). Then $u_n := 1-e_n \in D(\mathcal{E}^0_\varphi)$, $0 \le u_n \le 1$ and $u_n = 0$ $\varphi^2 \cdot dx$-a.e. on K_n^c. Moreover, by Proposition 3.1 and [MR 92, Chap.III, Proposition 2.12] $u_n \underset{n\to\infty}{\longrightarrow} 1$ w.r.t. $\mathcal{E}^0_{\varphi,1} := \mathcal{E}^0_\varphi + (\,,\,)_{L^2(\mathbb{R}^d;\varphi^2 dx)}$. Set $\overline{u}_n := u\, u_n$. Clearly, $\overline{u}_n \underset{n\to\infty}{\longrightarrow} u$ in $L^2(\mathbb{R}^d; \varphi^2 \cdot dx)$, $\overline{u}_n \in D(\mathcal{E}^+_\varphi)$ and

$$\sup_n \mathcal{E}^+_\varphi(\overline{u}_n, \overline{u}_n)^{1/2} \le \mathcal{E}^+_\varphi(u,u)^{1/2} + \|u\|_\infty \sup_n \mathcal{E}^0_\varphi(u_n, u_n)^{1/2} < \infty$$

(e.g. by [MR 92, Chap.I, Corollary 4.15]). Again by the Banach-Saks theorem it is sufficient to show that $\overline{u}_n \in D(\mathcal{E}^0_\varphi)$ for every $n \in \mathbb{N}$. So fix $n \in \mathbb{N}$. Since $u \in H^{1,2}(\mathbb{R}^d; dx)$, there are $f_m \in C_0^\infty(\mathbb{R}^d)$, $m \in \mathbb{N}$, such that

$$\int \|\nabla f_m - \nabla u\|^2_{\mathbb{R}^d} dx + \int |f_m - u|^2 dx \to 0 \quad \text{as } m \to \infty$$

and

$$\sup_m \|f_m\|_\infty < +\infty\ .$$

We have that $\overline{f}_m := f_m u_n \in D(\mathcal{E}^0_\varphi)$, $\overline{f}_m \to \overline{u}_n$ in $L^2(\mathbb{R}^d; \varphi^2 dx)$ as $m \to \infty$, and

$$\begin{aligned}
\sup_m \mathcal{E}^0_\varphi(\overline{f}_m, \overline{f}_m) &= \sup_m \int \|\tilde{\nabla}(f_m u_n)\|^2_{\mathbb{R}^d}\ \varphi^2 dx \\
&\le 2\sup_m \int \|\nabla f_m\|^2_{\mathbb{R}^d}\ c_n^2\ dx + 2\sup_m \int \|\tilde{\nabla} u_n\|^2_{\mathbb{R}^d} |f_m|^2 \varphi^2 dx \\
&< +\infty
\end{aligned}$$

with $c_n := \sup_{K_n} \tilde{\varphi}^2$, where we used the product rule for $(\tilde{\nabla}, D(\mathcal{E}^0_\varphi)) :=$ closure of $(\nabla, \mathcal{F}C_b^\infty)$ on $L^2(\mathbb{R}^d; \varphi^2 dx)$. Another application of the Banach-Saks Theorem implies that

$$\overline{u}_n \in D(\mathcal{E}^0_\varphi).$$

This completes the proof. □

Now we prove Proposition 1.5. We first need

Lemma 3.2. *Suppose* $\rho \in H^{1,1}(\mathbb{R}^d; dx)$ *(=Sobolev space of order 1 in* $L^1(\mathbb{R}^d; dx)$*),* $\rho \geq 0$*, such that for* $1 \leq i \leq d$

$$\rho^{-1/2} \frac{\partial \rho}{\partial x_i} \subset L^2(\mathbb{R}^d; dx)$$

then $\varphi := \rho^{1/2} \in H^{1,2}(\mathbb{R}^d; dx)$ *and* $\frac{\partial \varphi}{\partial x_i} = \frac{1}{2}\rho^{-1/2}\frac{\partial \rho}{\partial x_i}$, $1 \leq i \leq d$. *(Here and henceforth we set* $a/0 := sign(a)\cdot(+\infty)$ *for* $a \in \mathbb{R}$ *and* $(\pm\infty)\cdot 0 = 0$*).*

Proof. Let $n \in \mathbb{N}$ and $\varphi_n := \sqrt{\rho + 1/n}$. Then $\varphi_n \in H^{1,2}_{loc}(\mathbb{R}^d; dx)$ and for $1 \leq i \leq d$

$$\frac{\partial \varphi_n}{\partial x_i} = 2^{-1}(\rho + 1/n)^{-1/2}\frac{\partial \rho}{\partial x_i}$$

which converges to $2^{-1}\rho^{-1/2}\frac{\partial \rho}{\partial x_i}$ in $L^2(\mathbb{R}^d; dx)$ as $n \to \infty$. Hence the assertion follows. □

Proof of Proposition 1.5. By assumption it follows that the forms

$$\mathcal{E}_i(u, v) := \int \frac{\partial u}{\partial x_i}\frac{\partial v}{\partial x_i} d\nu; \quad u, v \in C_0^\infty(\mathbb{R}^d)$$

are closable on $L^2(\mathbb{R}^d; \nu)$ for every $1 \leq i \leq d$. Hence by Theorem 5.2 in [AR 90] (= Proof of the "Fukushima conjecture") there exists $\rho \in L^1(\mathbb{R}^d; dx)$, $\rho \geq 0$, such that $\nu = \rho \cdot dx$. (The existence of ρ also follows from [Ma 78, Part I, §1].) Using the assumption again it follows that $\rho \in H^{1,1}(\mathbb{R}^d; dx)$ with $\frac{\partial \rho}{\partial x_i} = \beta_i \rho$ $(\in L^1(\mathbb{R}^d; dx))$ and hence that

$$\rho^{-1}\frac{\partial \rho}{\partial x_i} = \beta_i \in L^2(\mathbb{R}^d; \rho dx) .$$

Since $\frac{\partial \rho}{\partial x_i} = 0$ dx-a.e. on $\{\rho = 0\}$ we deduce that $\rho^{-1/2}\frac{\partial \rho}{\partial x_i} \in L^2(\mathbb{R}^d; dx)$, $1 \leq i \leq d$. Lemma 3.2 now implies the assertion. □

4 PROOFS FOR THE RESULTS IN INFINITE DIMENSIONS

Consider the situation of Sect. 2. Before we prove Theorem 1.4 we need some preparations. Fix linear independent $k_1, \ldots, k_N \in K$ $(\subset E' \subset H \subset E)$, $N \in \mathbb{N}$, and we identify their linear span with $\mathbb{R}^N$. Let E_N be a closed linear subspace of E such that $E = \mathbb{R}^N \oplus E_N$. Let P_N, π_N denote the canonical projections onto $\mathbb{R}^N$, E_N respectively. The measure μ disintegrates w.r.t. π_N, i.e., there exists a kernel $\rho_N(\cdot, dx)$: $E_N \times \mathcal{B}(\mathbb{R}^N) \to [0,1]$ such that

$$\mu(dz) = \rho_N(y, dx)\nu_N(dy) \tag{4.1}$$

where $x = (x_1, \ldots, x_N) \in \mathbb{R}^N$, $z = \sum_{i=1}^N x_i k_i \oplus y$, $y \in E_N$, and $\nu_N = \mu \circ \pi_N^{-1}$ (i.e., the image measure of μ under π_N). Note that as in (2.1) we have that

$$L^2(E;\mu) = \int^{\oplus} L^2(\mathbb{R}^N; \rho_N(y, dx))\nu_N(dy)$$

and every $u \in L^2(E;\mu)$ corresponds to a measurable field of vectors (w.r.t. μ) $(u_y)_{y \in E_N}$ where $u_y := u(\cdot + y)$, $y \in E_N$.

Proposition 4.1. *For ν_N-a.e. $y \in E_N$, there exists $\varphi_N(\cdot, y) \in H^{1,2}(\mathbb{R}^N; dx)$ such that*

$$\rho_N(y, dx) = \varphi_N^2(x, y)dx$$

Proof. Since E_N is a separable Banach space the set $\mathcal{L}$ of all functions $T : E_N \to \mathbb{R}$ of the form $T(y) = f(l_1(y), \ldots, l_m(y))$, $y \in E_N$, $m \in \mathbb{N}$, $f \in C_b^\infty(\mathbb{R}^m)$, $l_i \in E_N'$, $1 \le i \le m$, is dense in $L^2(E_N; \nu_N)$. For all $T \in \mathcal{L}$, $g \in C_0^\infty(\mathbb{R}^N)$, and $1 \le i \le N$, by (4.1) and assumption (1.5) we have for $v := (T \circ \pi_N)(g \circ P_N)$

$$\begin{aligned}
&\int_{E_N} T(y) \int_{\mathbb{R}^N} \frac{\partial g}{\partial x_i} \rho_N(y, dx)\nu_N(dy) \\
&= \int \frac{\partial v}{\partial k_i} d\mu = -\int v \beta_{k_i} d\mu \\
&= \int_{E_N} T(y) \int g(x) \beta_{k_i}(x + y)\rho_N(y, dx)\nu_N(dy) .
\end{aligned}$$

Since $\beta_{k_i}(\cdot + y) \in L^2(\mathbb{R}^N; \rho_N(y, dx))$ for ν_N-a.e. $y \in E_N$ the assertion now follows from Proposition 1.5. □

Note that by [Dy 72, Lemma 4.1] φ_N in Proposition 4.1 can be assumed to be $\mathcal{B}(\mathbb{R}^N) \otimes \mathcal{B}(E_N)$-measurable. Then for $\varphi := [\int \varphi_N^2(\cdot, y)\nu_N(dy)]^{1/2}$ clearly,

$$\mu \circ P_N^{-1} = \varphi^2 \cdot dx . \tag{4.2}$$

Lemma 4.2. *Let φ be as in (4.2). Then $\varphi \in H^{1,2}(\mathbb{R}^N; dx)$ and $2\varphi^{-1} \frac{\partial \varphi}{\partial x_i} = \varphi^{-2} \int_{E_N} \beta_{k_i}(\cdot + y)\varphi_N^2(\cdot, y)\nu_N(dy)$, $1 \le i \le N$.*

Proof. We set $\rho := \varphi^2$. By Fubini's theorem and (1.5), for $g \in C_0^\infty(\mathbb{R}^N)$ and $1 \le i \le N$ we have that

$$\begin{aligned}\int_{\mathbb{R}^N} \frac{\partial g}{\partial x_i}\rho dx &= \int_{E_N}\int_{\mathbb{R}^N} \frac{\partial g}{\partial x_i}\varphi_N^2(x,y)dx\nu_N(dy) \\ &= \int \frac{\partial(g\circ P_N)}{\partial k_i}d\mu = -\int g\circ P_N\ \beta_{k_i}d\mu \\ &= -\int_{\mathbb{R}^N} g\int_{E_N}\beta_{k_i}(x+y)\varphi_N^2(x,y)\nu_N(dy)dx\ .\end{aligned}$$

Hence $\rho \in H^{1,1}(\mathbb{R}^N; dx)$ and $\frac{\partial\rho}{\partial x_i} = \int_{E_N}\beta_{k_i}(\cdot+y)\varphi_N^2(\cdot,y)\nu_N(dy)$, $1 \le i \le N$. Furthermore, by the Cauchy-Schwarz inequality (applied to the measure $\varphi_N^2(x,y)\nu_N(dy)$) for $1 \le i \le N$

$$\begin{aligned}\rho^{-1/2}\left|\frac{\partial\rho}{\partial x_i}\right| &\le \rho^{-1/2}\Big(\int_{E_N}\beta_{k_i}^2(\cdot+y)\varphi_N^2(\cdot,y)\nu_N(dy)\Big)^{1/2} \\ &\quad\cdot\Big(\int_{E_N}\varphi_N^2(\cdot,y)\nu_N(dy)\Big)^{1/2} \\ &= \Big(\int_{E_N}\beta_{k_i}^2(\cdot+y)\varphi_N^2(\cdot,y)\nu_N(dy)\Big)^{1/2} \in L^2(\mathbb{R}^N;dx)\end{aligned}$$

and the assertion follows by Lemma 3.2. □

Proof of Theorem 1.4. By Theorem 2.3 we have to show that $\mathcal{F}C_b^\infty$ is dense in $D(\mathcal{E}_{\mu,H,K})$ w.r.t. $\mathcal{E}_{\mu,H,K} + (\ ,\ )_{L^2(E;\mu)}$. Fix $N \in \mathbb{N}$ and let F_N be the set of all $\mathcal{B}_N$-measurable, bounded $u \in L^2(E;\mu)$ such that for every $1 \le i \le N$ there exists $f_i \in L^2(E;\mu)$ satisfying

$$\int \frac{\partial}{\partial k_i}(g\circ P_N)\,u\ d\mu = -\int (g\circ P_N)(f_i + u\beta_{k_i})d\mu \quad \text{for all } g \in C_0^\infty(\mathbb{R}^N). \tag{4.2a}$$

Claim. $F_N \subset D(\mathcal{E}^0_{\mu,H})$, $\frac{\partial u}{\partial k_i} = E_\mu[f_i\,|\,\mathcal{B}_N]$ if $i \le N$ and $\frac{\partial u}{\partial k_i} = 0$ if $i \ge N+1$ for all $u \in F_N$ with f_i as in (4.2a).

Let $u = h\circ P_N \in F_N$ for some $h \in L^2(\mathbb{R}^N; \varphi^2dx)$ with φ as in (4.2). We may asssume that u and hence that h is bounded. Let $v := g\circ P_N \in \mathcal{F}C_b^\infty$ with $g \in C_0^\infty(\mathbb{R}^N)$. Then for all $1 \le i \le N$ by (4.2)

$$\int_{\mathbb{R}^N}\frac{\partial g}{\partial x_i}h\varphi^2dx$$

$$\begin{aligned}
&= \int \frac{\partial v}{\partial k_i} u d\mu \\
&= -\int v \frac{\partial u}{\partial k_i} d\mu - \int v u \beta_{k_i} d\mu \\
&= -\int_{\mathbb{R}^N} g(\varphi^{-2} \int \frac{\partial u}{\partial k_i}(\cdot + y)\varphi_N^2(\cdot, y)\nu_N(dy))\varphi^2 dx \\
&\quad - \int_{\mathbb{R}^N} gh(\varphi^{-2} \int \beta_{k_i}(\cdot + y)\varphi_N^2(\cdot, y)\nu_N(dy))\varphi^2 dx .
\end{aligned}$$

Since by (4.2) and the Cauchy-Schwarz inequality

$$\begin{aligned}
&\int_{\mathbb{R}^N} \varphi^{-4}[\int_{E_N} \frac{\partial u}{\partial k_i}(\cdot + y)\varphi_N^2(\cdot, y)\nu_N(dy)]^2 \varphi^2 dx \\
&\leq \int_{\mathbb{R}^N} \varphi^{-2}[\int_{E_N} (\frac{\partial u}{\partial k_i}(\cdot + y))^2 \varphi_N^2(\cdot, y)\nu_N(dy)] \\
&\qquad \cdot [\int_{E_N} \varphi_N^2(\cdot, y)\nu_N(dy)] dx < \infty ,
\end{aligned}$$

by Lemmas 4.2, 2.2, and equation (3.3) it follows that

$$h \in H^{1,2}(\mathbb{R}^N; \varphi^2 dx) .$$

Consequently, again by Lemma 4.2 and Theorem 1.1 there exist $u_n \in C_0^\infty(\mathbb{R}^N)$, $n \in \mathbb{N}$, such that $u_n \underset{n\to\infty}{\longrightarrow} h$ in $L^2(\mathbb{R}^N; \varphi^2 dx)$ (hence $u_n \circ P_N \underset{n\to\infty}{\longrightarrow} u$ in $L^2(E; \mu)$) and

$$\sum_{i=1}^N \int (\frac{\partial}{\partial x_i}(u_n - u_m))^2 \varphi^2 dx \underset{n,m\to\infty}{\longrightarrow} 0 .$$

But the left hand side is equal to

$$\begin{aligned}
&\sum_{i=1}^N \int (\frac{\partial}{\partial k_i}(u_n \circ P_N - u_m \circ P_N))^2 d\mu \\
&= \mathcal{E}^0_{\mu,H}(u_n \circ P_N - u_m \circ P_N, u_n \circ P_N - u_m \circ P_N).
\end{aligned}$$

By the closedness of $(\mathcal{E}_{\mu,H}, D(\mathcal{E}^0_{\mu,H}))$ the first part of the claim is proven. Furthermore, we have shown that $u_n \circ P_N \to u$ w.r.t. $\mathcal{E}^0_{\mu,H} + (\ ,\ )_{L^2(E;\mu)}$ as $n \to \infty$, which implies the last part of the claim and that for $i \leq N$, $\frac{\partial u}{\partial k_i}$ is $\mathcal{B}_N$-measurable. Since $\{g \circ P_N \mid g \in C_0^\infty(\mathbb{R}^N)\}$ generates $\mathcal{B}_N$, Lemma 2.2 and (4.2a) entails that $\frac{\partial u}{\partial k_i} = E[f_i \mid \mathcal{B}_N]$ for $i \leq N$, and the claim is proven completely.

Using the Claim and condition (1.7) we can now complete the proof. Let $u \in D(\mathcal{E}^+_{\mu,H,K})$. We may assume that u is bounded. Let $u_N := E_\mu[u \,|\, \mathcal{B}_N]$, $N \in \mathbb{N}$. Then for all $g \in C_0^\infty(\mathbb{R}^N)$ and $1 \leq i \leq N$

$$\begin{aligned}
\int \frac{\partial}{\partial k_i}(g \circ P_N)\, u_N \, d\mu &= \int \frac{\partial}{\partial k_i}(g \circ P_N)\, u \, d\mu \\
&= -\int g \circ P_N(\frac{\partial u}{\partial k_i} + u\beta_{k_i}) d\mu \\
&= -\int g \circ P_N(E_\mu[\frac{\partial u}{\partial k_i} \,|\, \mathcal{B}_N] + E_\mu[u\beta_{k_i} \,|\, \mathcal{B}_N]) d\mu \\
&= -\int g \circ P_N(E_\mu[\frac{\partial u}{\partial k_i} \,|\, \mathcal{B}_N] + \\
&\qquad E_\mu[u(\beta_{k_i} - E_\mu[\beta_{k_i} \,|\, \mathcal{B}_N]) \,|\, \mathcal{B}_N] + u_N \beta_{k_i}) d\mu \ ,
\end{aligned}$$

where we used Lemma 2.2 for the second equality. Hence $u_N \in F_N$ and for $1 \leq i \leq N$ by (the second part of) the Claim

$$\frac{\partial u_N}{\partial k_i} = E_\mu[\frac{\partial u}{\partial k_i} \,|\, \mathcal{B}_N] + E_\mu[u(\beta_{k_i} - E_\mu[\beta_{k_i} \,|\, \mathcal{B}_N]) \,|\, \mathcal{B}_N]$$

(see also [T 87, Lemma 4]). Furthermore, again by the Claim $u_N \in D(\mathcal{E}^0_{\mu,H})$ and

$$\begin{aligned}
\mathcal{E}^0_{\mu,H}(u_N, u_N) &= \sum_{i=1}^N \int (\frac{\partial u_N}{\partial k_i})^2 d\mu \\
&\leq 2\mathcal{E}^+_{\mu,H,K}(u,u) + 2\|u\|_\infty^2 \sum_{i=1}^N \|\beta_{k_i} - E_\mu[\beta_{k_i} \,|\, \mathcal{B}_N]\|^2_{L^2(E;\mu)}
\end{aligned}$$

Hence by (1.7) $\sup_N \mathcal{E}^0_{\mu,H}(u_N, u_N) < \infty$. Since $u_N \underset{n\to\infty}{\longrightarrow} u$ in $L^2(E;\mu)$, the Banach-Saks theorem implies that $u \in D(\mathcal{E}^0_{\mu,H})$ (cf. [MR 92, Chap.I, Lemma 2.12]). □

Proof of Corollary 1.6 By [RZ 92a, Proposition 2.1] it follows from the conditions on φ that all a priori assumptions in Theorem 1.4 are fulfilled with μ replaced by $\overline{\mu} := \varphi^2\mu$, β_k by $\overline{\beta}_k := \beta_k + 2\varphi^{-1}\frac{\partial\varphi}{\partial k}$, $k \in K$, and $S_{\mu,H,K}$ replaced by $S_{\mu,H,K} + 2\varphi^{-1}\langle\nabla\varphi, \nabla\cdot\rangle_H$. By Theorem 1.4 it remains to check (1.7), i.e.,

$$\sup_N \sum_{i=1}^N \|\overline{\beta}_{k_i} - E_{\overline{\mu}}[\overline{\beta}_{k_i} \,|\, \mathcal{B}_N]\|^2_{L^2(E;\overline{\mu})} < \infty \ .$$

But because of the special form of $\overline{\beta}_{k_i}$ and the assumption this holds since by Minkowski's and Jensen's inequality

$$\begin{aligned}
&\sup_N \sum_{i=1}^{N} \|\varphi^{-1}\frac{\partial\varphi}{\partial k_i} - E_{\overline{\mu}}[\varphi^{-1}\frac{\partial\varphi}{\partial k_i} \,|\, \mathcal{B}_N]\|^2_{L^2(E;\overline{\mu})} \\
&\leq 2\sup_N \sum_{i=1}^{N}[\|\frac{\partial\varphi}{\partial k_i}\|^2_{L^2(E;\mu)} + \|E_{\overline{\mu}}[(\varphi^{-1}\frac{\partial\varphi}{\partial k_i})^2 \,|\, \mathcal{B}_N]\|^2_{L^1(E;\overline{\mu})}] \\
&\leq 2\int \|\nabla\varphi\|^2_H d\mu < \infty \,.
\end{aligned}$$

□

Proof of Corollary 1.7. Since we are going to change the Hilbert space H and hence the embedding $E' \subset H' \equiv H \subset E$, to avoid confusion we denote this embedding of E' into $H(\subset E)$ by i. Let $l_n \in E'$, $n \in \mathbb{N}$, such that $k_n := i(l_n)$, $n \in \mathbb{N}$, form an orthonormal basis of H in K. Let for $n \in \mathbb{N}$

$$c_n := \int \beta^2_{k_n} d\mu$$

and $\lambda_n \geq 1$ such that $\sum_{n=1}^{\infty} c_n/\lambda_n < \infty$ (e.g. $\lambda_n := (c_n+1)n^2$). Define

$$\tilde{H} := \{h \in H \,|\, \sum_{n=1}^{\infty} \langle h, k_n\rangle^2_H \ \lambda_n < \infty\}$$

with inner product $\langle h_1, h_2\rangle_{\tilde{H}} := \sum_{n=1}^{\infty} \langle h_1, k_n\rangle_H \langle h_2, k_n\rangle_H \ \lambda_n$; $h_1, h_2 \in \tilde{H}$. Then $\tilde{H} \subset H$ densely and continuously. Let $\tilde{K}$ denote the linear span of $\{l_n \,|\, n \in \mathbb{N}\}$. If j denotes the embedding of E' into $\tilde{H}$ given by $E' \subset \tilde{H}' \equiv \tilde{H}$ $(\subset H \subset E)$ then since

$${}_{E'}\langle l_n, h\rangle_E = \langle k_n, h\rangle_H = \lambda_n^{-1}\langle k_n, h\rangle_{\tilde{H}}, \quad h \in \tilde{H},\ n \in \mathbb{N},$$

it follows that $j(l_n) = \lambda_n^{-1} k_n$, $n \in \mathbb{N}$, which is well-μ-admissible in E with $\beta_{j(l_n)} = \lambda_n^{-1}\beta_{k_n}$. Hence

$$\sum_{n=1}^{\infty} \int \beta^2_{j(l_n)} d\mu = \sum_{n=1}^{\infty} c_n/\lambda_n < \infty$$

and for this choice of $\tilde{H}, \tilde{K}$ it follows by Minkowski's and Jensen's inequality that (1.7) holds. Thus, the assertion follows by Theorem 1.4. □

Proof of Theorem 1.9. It has been shown in [AR 91] that the process $\mathbf{M}_{\mu,H}$ (introduced in Sect. 1) solves the martingale problem for $(S_{\mu,H,K}, \mathcal{F}C_b^\infty(K))$. Let $\mathbf{M} = (\Omega, \mathcal{F}, (\mathcal{F}_t), (X_t), (P_z)_{z\in E})$ be another μ-symmetric right process solving this martingale problem. By e.g. [MR 92, Chap.II, Sect.4] $\mathbf{M}$ is associated with a symmetric Dirichlet form $(\mathcal{E}, D(\mathcal{E}))$ on $L^2(E;\mu)$. Hence by [MR 92, Chap.IV, Theorem 6.4] it suffices to show that

$$(\mathcal{E}, D(\mathcal{E})) - (\mathcal{E}^0_{\mu,H}, D(\mathcal{E}^0_{\mu,H})) \ . \tag{4.3}$$

Let $(L, D(L))$ denote the generator of $(\mathcal{E}, D(\mathcal{E}))$ on $L^2(E;\mu)$ and $T_t := e^{tL(\mathcal{E})}$, $t > 0$ (cf. [MR 92, Chap.I]). Let $u \in \mathcal{F}C_b^\infty(K)$ and $v \in L^2(E;\mu) \cap L^\infty(E;\mu)$. Then if $E_\mu[\ ]$ denotes expectation w.r.t. $P_\mu := \int P_z \mu(dz)$,

$$\begin{aligned}
\frac{1}{t}(u - T_t u, v)_{L^2(E;\mu)} &= \frac{1}{t} E_\mu[(u(X_0) - u(X_t))v(X_0)] \\
&= -\frac{1}{t} E_\mu[\int_0^t S_{\mu,H,K} u(X_s) ds\ v(X_0)] \\
&= -\frac{1}{t} \int_0^t (T_s S_{\mu,H,K} u, v)_{L^2(E;\mu)} ds \\
&\longrightarrow_{t\downarrow 0} (-S_{\mu,H,K} u, v)_{L^2(E;\mu)} \ .
\end{aligned}$$

Therefore, setting $u = v$ we conclude that $\mathcal{F}C_b^\infty(K) \subset D(\mathcal{E})$ and that (cf. e.g. [MR 92, Chap.I, end of Sect.2])

$$\mathcal{E}(u, v) = (-S_{\mu,H,K} u, v) \quad \text{for all } u \in \mathcal{F}C_b^\infty(K), v \in D(\mathcal{E}) \ .$$

Hence $\mathcal{F}C_b^\infty(K) \subset D(L)$ and

$$L = S_{\mu,H,K} \ \text{ on } \mathcal{F}C_b^\infty(K)$$

(cf. e.g. [MR 92, Chap.I, Proposition 2.16]). By Markov uniqueness (4.3) follows and the proof is complete. □

Acknowledgement

The second named author would like to thank the organizers for the warm hospitality during a very nice and stimulating stay at the winter school in Friedrichroda. Part of this work was done during a very pleasant stay of the first and second named authors at the S.I.S.S.A. in Trieste. We would like to thank our hosts, in particular Gianfausto

Dell' Antonio, for the hospitality and financial support. We also thank Shigeo Kusuoka for early stimulating discussions on topics related to the present paper and a very thorough referee for a lot of useful comments. Financial support of the SFB's 237 and 256 in Bochum-Düsseldorf-Essen resp. Bonn and the S.E.R.C. (grant No.: ERH 08372) is gratefully acknowledged.

REFERENCES

[AH-K 77] Albeverio, S., Høegh-Krohn, R.: Dirichlet forms and diffusion processes on rigged Hilbert spaces. Z. Wahrsch. verw. Geb. **40**, 1-57 (1977).

[AKR 90] Albeverio, S., Kusuoka, S., Röckner, M.: On partial integration in infinite dimensional space and applications to Dirichlet forms. J. London Math. Soc. **42**, 122-136 (1990).

[AR 90] Albeverio, S., Röckner, M.: Classical Dirichlet forms on topological vector spaces – closability and a Cameron-Martin formula. J. Funct. Anal. **88**, 395-436 (1990).

[AR 91] Albeverio, S., Röckner, M.: Stochastic differential equations in infinite dimensions: solutions via Dirichlet forms. Probab. Th. Rel. Fields **89**, 347-386 (1991).

[ARZ 92] Albeverio, S., Röckner, M., Zhang, T.S.: Markov uniqueness and its applications to martingale problems, stochastic differential equations and stochastic quantization. Preprint (1992). To appear in C. R. Math. Rep. Acad. Sci. Canada.

[Dy 72] Dynkin, E.B.: Integral representation of excessive measures and excessive functions. Usbekhi Mat. Nauk. **27**, No.1 (1972), 43-80. English translation: Russian Math. Surveys **27**, No.1 (1972), 43-84.

[F 80] Fukushima, M.: Dirichlet forms and Markov processes. Amsterdam – Oxford – New York: North Holland (1980).

[IW 89] Ikeda, N., Watanabe, S.: Stochastic differential equations and diffusion processes. North-Holland/Kodansha: Amsterdam – Tokyo 1989.

[KZ 71] Kailath, T., Zakai, M.: Absolute continuity and Radon-Nikodym derivatives for certain measures. Ann. Math. Statistics **42**, 130-140 (1971).

[LS 77] Lipster, R.S., Shiryayev, A.N.: Statistics of Random Processes I. General Theory. Springer: New York 1977.

[Ma 78] Malliavin, P.: Stochastic calculus of variation and hypoelliptic operators. Proc. of the International Symposium on Stochastic Differential Equations Kyoto 1976, Tokyo 1978.

[Mi 73] Mizohata, S.: The theory of partial differential equations. London: Cambridge University Press 1973.

[MR 92] Ma, Z.M., Röckner, M.: An introduction to the theory of (non-symmetric) Dirichlet forms. Springer: Berlin 1992.

[R 90] Röckner, M.: Dirichlet forms on infinite dimensional state space and applications. Lectures held at Silivri Summer School. Preprint (1990), and in: Stochastic Analysis and Related Topics, pp.131-186, ed.: H. Körezlioğlu, A.S. Ustunel. Birkhäuser: Boston 1992.

[R 91] Röckner, M.: Potential theory on non-locally compact spaces via Dirichlet forms. In: Potential Theory, pp.107-126, ed.: M. Kishi, Walter de Gruyter & Co.: Berlin – New York 1991.

[RZ 92a] Röckner, M., Zhang, T.S.: Uniqueness of generalized Schrödinger operators and applications. J. Funct. Anal. **105**, 187-231 (1992).

[RZ 92b] Röckner, M., Zhang, T.S.: Uniqueness of generalized Schrödinger operators – Part II, Preprint 1992.

[Sch90] Schmuland, B.: An alternative compactification for classical Dirichlet forms on topological vector spaces. Stochastics **33**, 75-90 (1990).

[Sh 88] Sharpe, M.J.: General theory of Markov processes. New York: Academic Press 1988.

[T 85] Takeda, M.: On the uniqueness of the Markovian self-adjoint extension of diffusion operators on infinite dimensional space. Osaka J. Math. **22**, 733-742 (1985).

[T 87] Takeda, M.: On the uniqueness of the Markovian self-adjoint extension, in "Lect. Notes in Math. (Stochastic processes – Mathematics and Physics)", Vol.1250, 319-325, Berlin: Springer 1987.

[T 92] Takeda, M.: The maximum Markovian self-adjoint extensions of generalized Schrödinger operators. J. Math. Soc. Japan, **44**, 113-130 (1992).

A DYNAMIC PROGRAMMING APPROACH TO THE OPTIMALITY OF GENERAL GITTINS INDEX SEQUENCES[†]

NICOLE EL KAROUI
Laboratoire de Probabilités, Université Pierre et Marie Curie
4, place Jussieu - Tour 56, 75252 Paris Cedex 05, France

IOANNIS KARATZAS*
Department of Statistics, Columbia University
619 Mathematics Building, New York, N.Y. 10027, U.S.A.

Abstract We extend the dynamic programming approach of Whittle (1980) and Tsitsiklis (1986), to obtain a simple and constructive proof for the optimality of Gittins index processes in the general, non-Markovian dynamic allocation (or "multi-armed bandit") problem. This latter is formulated as a question of stochastic control with multi-parameter time, in the manner of Mandelbaum (1986). The approach yields an explicit expression for the value of this problem, and leads to various characterizations of Gittins indices. This paper is an expanded and more detailed version of El Karoui & Karatzas (1992).

1. INTRODUCTION

We consider in this paper the formulation, as a stochastic control problem for multi-parameter random sequences, of the general, non-markovian *dynamic allocation* (or multi-armed bandit) *problem* in discrete time: there are d independent "projects" (or "arms"), only one of which may be engaged (pulled) at any given time, while the others remain "frozen". By engaging a particular project one receives a certain random reward, at a rate depending on time and on the history

† Part of this work was carried out during a brief visit by the second author at the University Pierre et Marie Curie (Paris VI), under the NSF-INRIA Collaborative Research Effort in Stochastic Control Grant NSF-INT-89-06965.

* Research supported in part by the U.S. Army Research Office under Grant ARO-DAAL 03-91-G-0170.

of the project that is being engaged. The objective is to schedule the engagements of the various projects so, as to maximize total expected discounted reward over an infinite horizon.

Questions of this sort had been essentially open for a long time, at least since World War II, until John Gittins and his collaborators made fundamental contributions in the 70's that amounted to a breakthrough in this area (cf. the survey article Gittins (1979) and the book Gittins (1989)). Within a markovian framework for the evolution of the independent projects' states, Gittins demonstrated that it is possible to assign to each project an "index function" of its state, defined in terms of that project's dynamics only, and such that the following policy is optimal: *at any given time, compute the indices of different projects and engage a project with maximal index.* In his paper (1980) and book (1982), P. Whittle then provided elegant, insightful and mathematically concise demonstrations for the optimality of the Gittins rule. Whittle's work was later complemented by Tsitsiklis (1986), who established from first principles the completeness of the class of so-called "write-off" policies (cf. Proposition 5.2 below).

The work of Gittins, Whittle and Tsitsiklis was extended to a general, *non-markovian* setting, first by Varaiya et al. (1985), and then by Mandelbaum (1986) who also formulated the question of Dynamic Allocation as a control problem with time-parameter in a multi-dimensional, partially-ordered set (as in Mandelbaum & Vanderbei (1981)). The arguments and proofs in both these works, however, are quite lengthy and demanding.

We offer in this paper an approach to the general, discrete-time problem, which combines the powerful formulation of Mandelbaum (1986) with the dynamic programming methodology of Whittle (1980). It is based on a very detailed study of an appropriate family of optimal stopping problems and of the associated Gittins index sequences, using tools from martingale theory. One is then able to compute explicitly the value random field of the dynamic allocation problem, in the spirit of Whittle, in terms of the values of the auxiliary stopping problems. In work currently in progress, we are extending this martingale/dynamic programming methodology to the continuous-time case; there we obtain not just more elegant arguments for known facts, but actually new results as well, stronger than those that seen possible by other methods.

The paper is an expanded and more detailed version of El Karoui & Karatzas (1992). It is organized as follows: in section 2, we formulate and study a *family of optimal stopping problems*, which allows the introduction of the Gittins index sequences and the study of their properties (section 3), including the "smallest reward-for-stopping" and "forwards-induction" characterizations. The dynamic allocation problem is formulated in section 4 in the manner of Mandelbaum (1986), and is solved in sections 5 and 6 by extending to the non-markovian case the dynamic programming methodology of Whittle (1980) as complemented by Tsitsiklis (1986). Section 7 discusses a representation of the value that involves the so-called "index random field", and studies its implications. All the proofs are collected in section 8.

2. A FAMILY OF OPTIMAL STOPPING PROBLEMS

Consider a probability space $(\Omega, \mathcal{F}, P)$ and on it a random sequence $H = \{h(t)\}_{t=1}^{\infty}$ which takes values in $[0, \infty)$, satisfies

$$E\sum_{t=0}^{\infty} \beta^t h(t+1) < \infty \tag{2.1}$$

for some given *discount factor* $\beta\epsilon(0,1)$, and is predictable with respect to a given filtration $\{\mathcal{F}(t)\}_{t=0}^{\infty}$. We assume that $\mathcal{F}(0) = \{\phi, \Omega\}$, denote by $\mathcal{S}(\theta)$ the class of $\{\mathcal{F}(t)\}$-stopping times with values in $\{\theta, \theta+1, \ldots\} \cup \{+\infty\}$ for any $\theta\epsilon\mathbf{N}_0$, and consider the family of *optimal stopping* problems, indexed by $m\epsilon[0,\infty)$:

(2.2)

$$V(t;m) \triangleq \mathrm{esssup}_{\tau\epsilon\mathcal{S}(t)} E\Big[\sum_{u=t}^{\tau-1} \beta^{u-t} h(u+1) + m\beta^{\tau-t} \,\Big|\, \mathcal{F}(t)\Big], \quad t\epsilon\mathbf{N}_0 \, .$$

From standard theory on optimal stopping (e.g. Neveu (1975), Chapter VI; Chow, Robbins & Siegmund (1991)) we know, for any given $m\epsilon[0,\infty)$, that the sequence

$$Z(t;m) \triangleq \beta^t V(t;m) + \sum_{u=0}^{t-1} \beta^u h(u+1), \quad t\epsilon\mathbf{N}_0 \tag{2.3}$$

is the *Snell envelope* of (i.e., the smallest nonnegative supermartingale that dominates) the sequence

$$(2.4) \qquad Y(t;m) \triangleq \beta^t m + \sum_{u=0}^{t-1} \beta^u h(u+1), \quad t \epsilon \mathbf{N}_0;$$

that the stopping time

$$(2.5) \qquad \sigma_t(m) \equiv \sigma(t;m) \triangleq \inf\{\theta \geq t / V(\theta;m) = m\}$$

is optimal for the problem of (2.2); that the sequence

$$(2.6) \qquad \{Z(\theta \wedge \sigma_t(m);m), \mathcal{F}(\theta)\}_{\theta=t}^{\infty} \quad \text{is a martingale;}$$

and that the following *Dynamic Programming Equation* holds, a.s.:

$$(2.7) \; V(t;m) = \max\Big[m, h(t+1) + \beta E\{V(t+1;m)|\mathcal{F}(t)\}\Big], \quad \forall \, t \epsilon \mathbf{N}_0 \, .$$

In addition, for any given $m \epsilon [0,\infty), V(\cdot\ ;m)$ is the *smallest* of the $\{\mathcal{F}(t)\}$ - adapted random sequences $U(\cdot)$, that satisfy a.s.

$$(2.8) \quad U(t) \geq \max\Big[m, h(t+1) + \beta E\{U(t+1)|\mathcal{F}(t)\}\Big], \quad \forall \, t \epsilon \mathbf{N}_0 \, .$$

Based on this theory one can establish the following facts (proved in section 8), for any given $t \epsilon \mathbf{N}_0$.

2.1 Lemma: *The mapping* $m \mapsto \sigma(t;m)$ *is decreasing and right-continuous with* $\sigma(t;0+) = \infty$ *and* $\sigma(t;\infty) = t$, *a.s.*

2.2 Lemma: *The mapping* $m \mapsto V(t;m)$ *is convex and increasing, with right-hand derivative* $\frac{\partial^+}{\partial m} V(t;m) \triangleq \lim_{\delta \downarrow 0} \frac{1}{\delta}\Big[V(t;m+\delta) - V(t;m)\Big]$ *given, for any* $m \epsilon [0,\infty)$, *by*

$$(2.9) \qquad \frac{\partial^+}{\partial m} V(t;m) = E\Big[\beta^{\sigma(t;m)-t} \,\Big|\, \mathcal{F}(t)\Big], \quad a.s.$$

3. GITTINS INDEX SEQUENCES

Let us denote by $\mathcal{M}(t)$ the class of all positive, $\mathcal{F}(t)$-measurable random variables. By analogy with Whittle (1980), Varaiya et al. (1985), Mandelbaum (1986), we define the *Gittins index*

$$(3.1) \qquad M(t) \triangleq \text{essinf}\{X\epsilon\mathcal{M}(t)/V(t;X) = X, a.s.\}$$

at time $t\epsilon\mathbf{N}_0$. It can be shown that this $\mathcal{F}(t)$-measurable random variable admits also the *forwards induction* characterization

$$(3.2) \quad (1-\beta)M(t) = \text{esssup}_{\tau\epsilon\mathcal{S}(t+1)} \frac{E\left[\sum_{u=t}^{\tau-1} \beta^u h(u+1)|\mathcal{F}(t)\right]}{E\left[\sum_{u=t}^{\tau-1} \beta^u|\mathcal{F}(t)\right]}, \quad a.s.$$

as the maximum achievable "conditional expected discounted reward, over conditional expected discounted time", from period t onwards (proved in section 8). The notation

$$(3.3) \qquad \underline{M}(t,\theta) \triangleq \min_{t\le u\le\theta} M(u)\ , \qquad \underline{M}(\theta) \triangleq M(0,\theta)$$

will be used throughout for the lower envelope of the *Gittins index random sequence* $M = \{M(t), \mathcal{F}(t)\}_{t\epsilon\mathbf{N}_0}$. The equivalences

(3.4)

$$\sigma(t;m) > \theta \;\Leftrightarrow\; V(s;m) > m,\; s = t, t+1, \ldots, \theta \;\Leftrightarrow\; m < \underline{M}(t,\theta)\ ,$$

for $\theta \ge t$, lead then to the following expressions for the optimal stopping time of (2.5) in terms of the index sequence, and conversely:

$$(3.5) \quad \sigma(t;m) = \inf\{\theta \ge t/\underline{M}(t,\theta) \le m\} = \inf\{\theta \ge t/M(\theta) \le m\}\ ,$$

$$(3.6) \qquad \underline{M}(t,\theta) = \inf\{m \ge 0/\sigma(t;m) \le \theta\}\ .$$

3.1 Remark: The identification (3.5) has the following implications: for any $t\epsilon\mathbf{N}_0$, we have

(3.7)

$$E\Big[\sum_{\theta=\sigma_t(m)}^{\infty} \beta^\theta h(\theta+1)\;\Big|\;\mathcal{F}(t)\Big] = (1-\beta)\cdot E\Big[\sum_{\theta=\sigma_t(m)}^{\infty} \beta^\theta \underline{M}(t,\theta)\;\Big|\;\mathcal{F}(t)\Big]\ ,$$

(3.8) $$E\Big[\sum_{\theta=t}^{\infty}\beta^{\theta}h(\theta+1)\,\Big|\,\mathcal{F}(t)\Big] = (1-\beta)\cdot E\Big[\sum_{\theta=t}^{\infty}\beta^{\theta}\underline{M}(t,\theta)\,\Big|\,\mathcal{F}(t)\Big]\,,$$

(3.9) $$V(t;m) = (1-\beta)\cdot E\Big[\sum_{\theta=t}^{\infty}\beta^{\theta-t}(m\vee\underline{M}(t,\theta))\,\Big|\,\mathcal{F}(t)\Big]\,,$$

almost surely. The proofs are given in section 8. Note the significance of (3.8): in the computation of expected future total discounted reward, conditionally on $\mathcal{F}(t)$, the reward sequence $\{h(\theta+1)\}_{\theta=t}^{\infty}$ may be replaced by the *decreasing* sequence $\{(1-\beta)\underline{M}(t,\theta)\}_{\theta=t}^{\infty}$.

3.2 Remark: For any given $t\epsilon\mathbf{N}_0$, the function $m\mapsto\frac{\partial^+}{\partial m}V(t;m)$ is right-continuous and increasing; it is equal to zero at $m=0$, and equal to one for $m\geq M(t)$, a.s. These properties follow readily from (3.4) and Lemma 2.2.

Let us give a few examples (inspired by Whittle (1982)), before discussing the dynamic allocation problem (sections 4-7).

3.3 The "deteriorating" case: Suppose $\{h(t)\}_{t\epsilon\mathbf{N}}$ is a.s. decreasing; then we have $M(t)=h(t+1)/(1-\beta)$ from (3.2), and from (3.9): $V(t;m)=E[\sum_{\theta=t}^{\infty}\beta^{\theta-t}(m\vee h(\theta+1))|\mathcal{F}(t)]$, a.s.

It follows then, from this and from Remark 3.1, that we have a.s.

(3.10) $$V(t;m)=V^*(t;m)\ ,\quad\forall\, t\epsilon\mathbf{N}_0,\ m\geq 0\ ,\quad\text{where}$$

(3.11)
$$V^*(t;m)\triangleq \operatorname{esssup}_{\tau\epsilon\mathcal{S}(t)}E\Big[\sum_{\theta=t}^{\tau-1}\beta^{\theta-1}(1-\beta)\underline{M}(t,\theta)+m\beta^{\tau-t}\,\Big|\,\mathcal{F}(t)\Big]$$

is the analogue of (2.1) for the "deteriorating" reward random sequence $h^*(\theta+1)\triangleq(1-\beta)\underline{M}(t,\theta)\ ,\ \theta\geq t$.

3.4 The "improving" case: Suppose that $\{V(t;m)\}_{t\epsilon\mathbf{N}_0}$ is increasing (for this, it suffices that $\{h(t)\}_{t\epsilon\mathbf{N}}$ be increasing) for all $m\geq 0$, almost surely. Then from (3.1): $V(t+1;M(t)-\frac{1}{n})\geq V(t;M(t)-\frac{1}{n})>M(t)-\frac{1}{n}$ and thus $M(t+1)>M(t)-\frac{1}{n},\ \forall\, n\geq 1$ whence $M(t+1)\geq M(t)$, almost surely. In other words, $\{M(t)\}_{t\epsilon\mathbf{N}_0}$

is also increasing, (3.9) gives $V(t;m) = m \vee M(t)$, and we have from this and (3.8):

$$M(t) = V(t;0) = E\Big[\sum_{\theta=t}^{\infty} \beta^{\theta-t} h(\theta+1) \;\Big|\; \mathcal{F}(t)\Big] , \quad a.s.$$

3.5 The "climactic" case" Suppose there is an $\{\mathcal{F}(t)\}$-stopping time $\rho \epsilon \mathcal{S}(1)$ such that $V(\rho+j;m) = m, \forall\, m \geq 0,\ \forall\, j \epsilon \mathbf{N}_0$ and $V(t+1;m) \geq V(t;m),\ \forall\, 0 \leq m \leq K$ on $\{t+1 < \rho\}$, almost surely. In other words, "continuation is always optimal all the way up to ρ, and becomes pointless beyond ρ". Then, just as before, we see that $M(0) \leq M(1) \leq \ldots \leq M(\rho-1)$ and $M(\rho+j) = 0,\ \forall\, j \epsilon \mathbf{N}_0$, almost surely. Therefore, we have on $\{t \leq \rho - 1\} : \underline{M}(t,\theta) = M(t) 1_{\{t \leq \theta \leq \rho-1\}}, \sigma_t(0) = \rho$ from (3.5), $V(t;0) = E\Big[\sum_{\theta=t}^{\rho-1} \beta^{\theta-t} h(\theta+1) \;\Big|\; \mathcal{F}(t)\Big]$ from (3.8) and (3.7), $V(t;m) = (1-\beta)(m \vee M(t)) \cdot E\Big[\sum_{\theta=t}^{\rho-1} \beta^{\theta-t} \;\Big|\; \mathcal{F}(t)\Big]$ from (3.9), and thus

$$(1-\beta)M(t) = \frac{E\Big[\sum_{\theta=t}^{\rho-1} \beta^{\theta} h(\theta+1) \;\Big|\; \mathcal{F}(t)\Big]}{E\Big[\sum_{\theta=t}^{\rho-1} \beta^{\theta} \;\Big|\; \mathcal{F}(t)\Big]} 1_{\{t \leq \rho-1\}} , \quad a.s.$$

3.6 The Markovian case: Suppose that $h(t+1) = \eta(X(t))$, $t \epsilon \mathbf{N}_0$, where $\mathbf{X} = (\{X(t), \mathcal{F}(t)\}_{t \epsilon \mathbf{N}_0}, \{\mathbf{P}^x\}_{x \epsilon S})$ is a time-homogeneous Markov family on the measurable state-space $(S, \mathcal{B})$ and $\eta : S \to [0,\infty)$ a given measurable function: $P[X(t+1) \epsilon A | \mathcal{F}(t)] = \mathbf{P}^x(A) \Big|_{x = X(t)}$ a.s. for $t \epsilon \mathbf{N}_0, A \epsilon \mathcal{B}$. Then $V(t;m) = \nu(X(t);m)$ and $M(t) = \mu(X(t))$ for all $t \epsilon \mathbf{N}_0$; here the *optimal expected reward*

$$(2.2)' \qquad \nu(x;m) \triangleq \sup_{\tau \epsilon \mathcal{S}(0)} \mathbf{E}^x \Big[\sum_{u=0}^{\tau-1} \beta^u \eta(X(u)) + m\beta^\tau\Big] \; ; \; x \epsilon S,\ m \geq 0$$

satisfies the Dynamic Programming equation

$$(2.7)' \qquad \nu(x;m) = \max\Big[m,\ \eta(x) + \beta \int_S \nu(y;m) \mathbf{P}^x(dy)\Big] ,$$

and the *index function*

$$\mu(x) \triangleq \inf\{m \geq 0 / \nu(x;m) = m\} \tag{3.1'}$$

satisfies

$$(1-\beta)\cdot\mu(x) = \sup_{\tau\epsilon\mathcal{S}(1)} \frac{\mathbf{E}^x \sum_{u=0}^{\tau-1} \beta^u \eta(X(u))}{\mathbf{E}^x \sum_{u=0}^{\tau-1} \beta^u}\ , \quad x\epsilon S\ . \tag{3.2'}$$

4. THE DYNAMIC ALLOCATION PROBLEM

Let us formulate now the Dynamic Allocation (or Multi-Armed Bandit) problem, in the manner of Mandelbaum (1986). Consider d *independent* filtrations $\mathbf{F}_i = \{\mathcal{F}_i(t)\}_{t\epsilon\mathbf{N}_0}, i = 1,\ldots,d$, and construct a new filtration $\mathbf{F} = \{\mathcal{F}(\underset{\sim}{s})\}_{\underset{\sim}{s}\epsilon S}$ by setting

$$\mathcal{F}(\underset{\sim}{s}) \triangleq \bigvee_{i=1}^{d} \mathcal{F}_i(s_i)\ , \quad \underset{\sim}{s} = (s_1,\ldots,s_d)\ \epsilon\ S \tag{4.1}$$

on the lattice $S = \mathbf{N}_0^d$ of nonnegative integers (endowed with the partial ordering $\underset{\sim}{r} \leq \underset{\sim}{s} \Leftrightarrow r_i \leq s_i,\ \ \forall\, i = 1,\ldots,d$). For any given $\underset{\sim}{s}\epsilon S$, an *allocation strategy* (or optional increasing path) *emanating from* $\underset{\sim}{s}$ is a random sequence $\underset{\sim}{T} = \{\underset{\sim}{T}(t)\}_{t\epsilon\mathbf{N}_0}$ with values in S, and such that

$$\begin{aligned}
&(i) && \underset{\sim}{T}(0) = \underset{\sim}{s}\\
&(ii) && \underset{\sim}{T}(t+1) = \underset{\sim}{T}(t) + \underset{\sim}{e}_i \quad \text{for some} \quad i = 1,\ldots,d, \quad \text{and}\\
&(iii) && \{\underset{\sim}{T}(t+1) = \underset{\sim}{T}(t) + \underset{\sim}{e}_i,\ \underset{\sim}{T}(t) = \underset{\sim}{r}\}\ \epsilon\ \mathcal{F}(\underset{\sim}{r}),\\
& && \qquad \forall\, i = 1,\ldots,d, \qquad \forall\, \underset{\sim}{r}\epsilon S\ .
\end{aligned} \tag{4.2}$$

Here, $\underset{\sim}{e}_i$ stands for the ith unit vector in S. We denote by $\mathcal{A}(\underset{\sim}{s})$ the class of such S-valued random sequences. Observe that, with $|\underset{\sim}{r}| \triangleq \sum_{i=1}^{d} r_i$ for $\underset{\sim}{r}\epsilon S$, we have $|\underset{\sim}{T}(t)| = t + |\underset{\sim}{s}|$ for every $\underset{\sim}{T}\epsilon\mathcal{A}(\underset{\sim}{s}), t\epsilon\mathbf{N}_0$.

Intuitively, $T_i(t)$ counts "how many times the ith arm has been pulled up to time t", and the sequence $\{T_i(t)\}_{t\epsilon\mathbf{N}_0}$ is nondecreasing. The requirements (4.2) express that $\underset{\sim}{T}$ "pulls one arm at a time", and that the decision about which arm to pull at time $t+1$ is to be made on the basis of the information accumulated from the pulls of the various arms up to time t.

Associated with every arm (or project) $i = 1, \ldots, d$ is a random sequence $H_i = \{h_i(t)\}_{t \in \mathbf{N}}$, which represents the "reward" obtained by pulling that particular arm (or engaging that particular project); it is predictable with respect to the filtration $\mathbf{F}_i$, and takes values in $[0, K(1-\beta)]$ for some fixed $K \epsilon (0, \infty)$. The *dynamic allocation* (or multi-armed bandit) *problem* consists of maximizing the expected reward $E[\mathcal{R}(\underset{\sim}{T})|\mathcal{F}(\underset{\sim}{s})]$ over all $\underset{\sim}{T} \epsilon \mathcal{A}(\underset{\sim}{s})$, where

$$(4.3) \qquad \mathcal{R}(\underset{\sim}{T}) \triangleq \sum_{t=0}^{\infty} \sum_{i=1}^{d} \beta^t h_i(T_i(t+1))[T_i(t+1) - T_i(t)] \,.$$

We denote the *value random field* of this problem by

$$(4.4) \qquad \Phi(\underset{\sim}{s}) \triangleq \operatorname{esssup}_{\underset{\sim}{T} \epsilon \mathcal{A}(\underset{\sim}{s})} E[\mathcal{R}(\underset{\sim}{T})|\mathcal{F}(\underset{\sim}{s})], \quad \underset{\sim}{s} \epsilon S \,,$$

and expect it to satisfy the formal *Bellman equation of Dynamic Programming* in this setup:

$$(4.5) \qquad \Phi(\underset{\sim}{s}) = \max_{1 \le j \le d} \left[h_j(s_j + 1) + \beta \cdot E\{\Phi(\underset{\sim}{s} + \underset{\sim}{e}_j)|\mathcal{F}(\underset{\sim}{s})\} \right], \quad \underset{\sim}{s} \epsilon S$$

(cf. Mandelbaum & Vanderbei (1981)). We shall verify this equation, and produce a rather explicit representation for $\Phi(\underset{\sim}{s})$ which will in turn enable us to discern an optimal allocation policy, following the method of Whittle (1980); cf. Corollaries 6.3 and 6.4. This method embeds the optimization problem of (4.4) into a family of problems of the same sort, but with the additional option of "retiring" (i.e., abandoning all projects) and receiving a reward $M \ge 0$; cf. (4.6) and (4.7). This parametrization is the same as in the family (2.2) of optimal stopping problems.

In order to carry out this embedding we shall need to extend the notion of an allocation strategy, to accommodate a stopping time. We say that a measurable function $\underset{\sim}{\nu} : \Omega \to S$ is a *stopping point* of $\mathbf{F}$, if $\{\underset{\sim}{\nu} = \underset{\sim}{s}\} \epsilon \mathcal{F}(\underset{\sim}{s})$, $\forall\, \underset{\sim}{s} \epsilon S$; and for any such $\underset{\sim}{\nu}$ we consider the σ-field

$$\mathcal{F}(\underset{\sim}{\nu}) \triangleq \{A \epsilon \mathcal{F} \ / \ A \cap \{\underset{\sim}{\nu} = \underset{\sim}{s}\} \ \epsilon \ \mathcal{F}(\underset{\sim}{s}), \ \ \forall\, \underset{\sim}{s} \epsilon S\} \,.$$

From the definition (4.2) of an allocation strategy, it follows that $\underset{\sim}{T}(t)$ is a stopping point for every $t \epsilon \mathbf{N}_0$; thus $\mathcal{G}(t) = \mathcal{F}(\underset{\sim}{T}(t))$ is defined, and $\mathbf{G} = \{\mathcal{G}(t)\}_{t \epsilon \mathbf{N}_0}$ is a filtration. A *policy* $\Pi = (\underset{\sim}{T}, \tau)$ is a pair consisting of an allocation strategy $\underset{\sim}{T} \epsilon \mathcal{A}(s)$, and of a stopping time τ of $\mathbf{G}$; we denote by $\mathcal{P}(\underset{\sim}{s})$ the class of such policies, and introduce the analogues of (4.3) - (4.5):

$$(4.6)\quad \mathcal{R}(\Pi; M) \triangleq \sum_{t=0}^{\tau-1} \sum_{i=1}^{d} \beta^t h_i(T_i(t+1))[T_i(t+1) - T_i(t)] + M\beta^\tau$$

$$(4.7)\qquad \Phi(\underset{\sim}{s}; M) \triangleq \mathrm{esssup}_{\Pi\epsilon\mathcal{P}(\underset{\sim}{s})} E[\mathcal{R}(\Pi; M)|\mathcal{F}(\underset{\sim}{s})], \quad \underset{\sim}{s}\epsilon S$$

(4.8)
$$\Phi(\underset{\sim}{s}; M) = \max\Big[M, \max_{1\le j\le d}\{h_j(s_j+1) + \beta \cdot E[\Phi(\underset{\sim}{s} + \underset{\sim}{e}_j; M)|\mathcal{F}(\underset{\sim}{s})]\}\Big] .$$

Furthermore, for any given $M\epsilon[0,\infty)$ the random field $\Phi(\cdot; M)$ of (4.7) is the *smallest* of the $\mathbf{F}$-adapted random fields $\Psi(\cdot)$ that satisfy a.s.
(4.9)
$$\Psi(\underset{\sim}{s}) \ge \max\big[M, \max_{1\le j\le d}\{h_j(s_j+1) + \beta . E[\Psi(\underset{\sim}{s} + \underset{\sim}{e}_j)|\mathcal{F}(\underset{\sim}{s})]\}\big] , \quad \forall\ \underset{\sim}{s}\epsilon S .$$

4.1 Remarks: For every $i = 1, \ldots, d$ we may consider the family of optimal stopping problems
(2.2)(i)
$$V_i(t; m) = \mathrm{esssup}_{\tau\epsilon\mathcal{S}_i(t)} E\Big[\sum_{u=0}^{\tau-1} \beta^{u-t} h_i(u+1) + m\beta^{\tau-t} \mid \mathcal{F}_i(t)\Big] , \quad t\epsilon\mathbf{N}_0$$

for the filtration $\mathbf{F}_i$ (parametrized by $m \ge 0$), as well as the associated Gittins index sequence $\{M_i(t)\}_{t\epsilon\mathbf{N}_0}$ of (3.1), which now takes values in $[0, K]$. All the results of sections 2 and 3 are in force for these new objects.

The Dynamic Programming equation (4.8) extends the equation (2.7) to the multi-dimensional case $d \ge 2$, and (4.9) is the analogue of (2.8).

5. CHARACTERIZATION OF AN OPTIMAL STOPPING TIME FOR (4.7)

For any fixed $i \epsilon \{1, \ldots, d\}$, introduce the notation $\underset{\sim}{s}^{(i)} = (s_1, \ldots, s_{i-1}, s_{i+1}, \ldots, s_d)$ for $\underset{\sim}{s} \epsilon S$, and consider the filtration $\mathbf{F}^{(i)} = \{\mathcal{F}^{(i)}(\underset{\sim}{s}^{(i)})\}_{\underset{\sim}{s}^{(i)} \epsilon \mathbf{N}_0^{d-1}}$ given by $\mathcal{F}^{(i)}(\underset{\sim}{s}^{(i)}) \triangleq \bigvee_{\substack{j=1 \\ j \neq i}}^{d} \mathcal{F}_j(s_j)$; furthermore, let $\Psi^{(i)}(\underset{\sim}{s}^{(i)}; M)$ denote the value function of a dynamic allocation problem which is just the same as (4.7) but *with the ith project deleted.* By analogy with (4.8), $\Psi^{(i)}(\underset{\sim}{s}^{(i)}; M)$ is the smallest among the $\mathbf{F}^{(i)}$ - adapted random fields $\Psi^{(i)}(\cdot)$ that satisfy

(5.1)

$$\Psi^{(i)}(\underset{\sim}{s}^{(i)}) \geq \max[M, \max_{j \neq i}\{h_j(s_j + 1) + \beta \cdot E[\Psi^{(i)}(\underset{\sim}{s}^{(i)} + \underset{\sim}{e}_j) | \mathcal{F}^{(i)}(\underset{\sim}{s}^{(i)})]\}]$$

$\forall\ \underset{\sim}{s}^{(i)} \epsilon \mathbf{N}_0^{d-1}$, almost surely. With this notation, the following Lemma and Proposition generalize results of Tsitsiklis (1986); they are proved in section 8.

5.1 Lemma: *For every* $i \epsilon \{1, \ldots, d\}$, $M \epsilon (0, \infty)$, $\underset{\sim}{s} \in S$, *we have a.s.*

(5.2)

$$\Psi^{(i)}(\underset{\sim}{s}^{(i)}; M) \vee V_i(s_i; M) \leq \Phi(\underset{\sim}{s}; M) \leq \Psi^{(i)}(\underset{\sim}{s}^{(i)}; M) + V_i(s_i; M) - M\ .$$

For any given $M \epsilon (0, \infty)$, consider now the subclass $\mathcal{P}^*(\underset{\sim}{s}, M)$ of $\mathcal{P}(\underset{\sim}{s})$ that consists of policies $\Pi = (\underset{\sim}{T}, \tau)$ with the following properties:

(a) Any project i is abandoned, as soon as $V_i(T_i(t); M) = M$; i.e., as soon as the allocation strategy $\underset{\sim}{T}(\cdot)$ has engaged it for $\sigma_i(s_i; M) - s_i$ times.

(b) One retires as soon as all the projects have been abandoned, and only then.

(c) Before retiring, one engages projects only from among those that have not yet been abandoned.

These are the so-called "*write-off*" policies of Whittle (1980). In particular, in the notation $\underset{\sim}{\sigma}(\underset{\sim}{s}; M) = \{\sigma_i(s_i; M)\}_{i=1}^{d}$, we have for any $\Pi = (\underset{\sim}{T}, \tau)\ \epsilon\ \mathcal{P}^*(\underset{\sim}{s}, M)$ of them: $\tau = \tau^*(\underset{\sim}{s}, M)$, where

$$\tau^*(\underset{\sim}{s}, M) \triangleq \sum_{i=1}^{d} [\sigma_i(s_i; M) - s_i] = |\underset{\sim}{\sigma}(\underset{\sim}{s}; M) - \underset{\sim}{s}|\ . \tag{5.3}$$

5.2 Proposition: *The class $\mathcal{P}^*(\underset{\sim}{s}, M)$ is complete for the problem of (4.7); more precisely, for an arbitrary policy $(\underset{\sim}{T}, \tau)\epsilon\mathcal{P}(\underset{\sim}{s})$, there is a "write-off" type policy $(\underset{\sim}{T}', \tau^*) \epsilon \mathcal{P}^*(\underset{\sim}{s}, M)$ which satisfies*

$$E[\mathcal{R}(\underset{\sim}{T}, \tau; M)|\mathcal{F}(\underset{\sim}{s})] \leq E[\mathcal{R}(\underset{\sim}{T}', \tau^*; M)|\mathcal{F}(\underset{\sim}{s})], \quad a.s. \tag{5.4}$$

in the notation of (4.6), where $\tau^* = \tau^*(\underset{\sim}{s}, M)$ is the $\{\mathcal{F}(\underset{\sim}{T}(t))\}_{t\epsilon\mathbf{N}_0}$ - stopping time of (5.3). In particular,

$$\Phi(\underset{\sim}{s}; M) = \text{esssup}_{\Pi\epsilon\mathcal{P}^*(\underset{\sim}{s},M)} E[\mathcal{R}(\Pi; M)|\mathcal{F}(\underset{\sim}{s})] \ . \tag{5.5}$$

Proposition 5.2 is the key step to the solution of the Dynamic Allocation problem, as we shall see in the next section; in particular, it provides a *separation principle* (between allocation and retirement) *for the problem of (4.7).*

6. THE WHITTLE REDUCTION

With the help of Proposition 5.2, we can now obtain an extension of Lemma 2.2 to the multi-dimensional case (Proposition 6.1), which in turn leads to the Whittle (1980) representation of the value random field (Theorem 6.2). Both these results, as well as their Corollaries, are proved in section 8.

6.1 Proposition: *The mapping $M \mapsto \Phi(\underset{\sim}{s}; M)$ is convex and increasing, with derivative given by*

$$\frac{\partial}{\partial M}\Phi(\underset{\sim}{s}; M) = \prod_{i=1}^{d}\Big(\frac{\partial^+}{\partial m}V_i(s_i; M)\Big) \ , \quad a.s. \tag{6.1}$$

for Lebesgue - a.e. $M\epsilon(0, \infty)$.

6.2 Theorem: *The value random field $\Phi(\underset{\sim}{s}; M)$ of (4.7) is given as*

(6.2)
$$\Phi(\underset{\sim}{s}; M) = \left\{ \begin{array}{lcr} M & ; & M \geq K \\ K - \int_M^K \Big(\prod_{i=1}^d \frac{\partial^+}{\partial m} V_i(s_i; m)\Big) dm & ; & 0 \leq M < K \end{array} \right\}$$

on S, and satisfies the Bellman equation (4.8); more precisely, we have

(6.3) $$\Phi(\underset{\sim}{s}; M) \geq M; \quad \forall\, M \geq 0\,, \quad \underset{\sim}{s} \epsilon S,$$

(6.4)
$$\Phi(\underset{\sim}{s}; M) \geq \max_{1 \leq j \leq d} \Big[h_j(s_j+1) + \beta \cdot E[\Phi(\underset{\sim}{s} + \underset{\sim}{e}_j; M) | \mathcal{F}(\underset{\sim}{s})\Big]; \quad \forall\, M \geq 0, \underset{\sim}{s} \epsilon S,$$

(6.5) $$\Phi(\underset{\sim}{s}; M) = M \quad on \quad \{M \geq \max_{1 \leq j \leq d} M_j(s_j)\}; \quad \forall\, \underset{\sim}{s} \epsilon S,$$

(6.6) $$\begin{aligned} \Phi(\underset{\sim}{s}; M) &= h_i(s_i + 1) + \beta \cdot E[\Phi(\underset{\sim}{s} + \underset{\sim}{e}_i; M) | \mathcal{F}(\underset{\sim}{s})] \\ &\quad on \quad \{M < M_i(s_i) = \max_{1 \leq j \leq d} M_j(s_j)\}, \quad \forall\, \underset{\sim}{s} \epsilon S. \end{aligned}$$

6.3 Corollary: *The value random field $\Phi(\underset{\sim}{s})$ of (4.4) is given as*

(6.7) $$\Phi(\underset{\sim}{s}) = K - \int_0^K \Big(\prod_{i=1}^d \frac{\partial^+}{\partial m} V_i(s_i; m) \Big) dm\,, \quad \underset{\sim}{s} \epsilon S\,,$$

and satisfies the Bellman equation (4.5).

6.4 Corollary: *Every allocation strategy $\underset{\sim}{T}^*$ of the "index-type," i.e., such that*

(6.8) $$\begin{aligned} M_i(T_i^*(t)) &= \max_{1 \leq j \leq d} M_j(T_j^*(t)) \\ \text{on } \{\underset{\sim}{T}^*(t+1) &= \underset{\sim}{T}^*(t) + \underset{\sim}{e}_i\}\,, \quad \forall\, i = 1, \ldots, d, \quad t \epsilon \mathbf{N}_0\,, \end{aligned}$$

is optimal for the problem of (4.4).

6.5 Remark: The significance of (6.5), (6.6) for the problem of (4.7) is the following:

(i) According to (6.6), it is optimal "always to pull an arm with maximal index, as long as this latter exceeds M".

(ii) According to (6.5), "all arms should be abandoned once all indices have fallen below (or at) the retirement reward M".

6.6 Remark: From (6.1) and (3.10), (3.11), it follows now that $\Phi(\underset{\sim}{0}) = \Phi^*(\underset{\sim}{0})$, where $\Phi^*(\underset{\sim}{0})$ is the value of the problem (4.4) with $\underset{\sim}{s} = \underset{\sim}{0}$ and with the reward sequences $h_i(t+1)$ replaced by the *decreasing* sequences $h_i^*(t+1) \triangleq (1-\beta)\underline{M}_i(t), t\epsilon \mathbf{N}_0$, for $i = 1, \dots, d$.

6.7 Remark: The results of section 5 are, strictly speaking, not necessary for our development; the arguments of this section show that the random field $F(\underset{\sim}{s}; M)$, defined as the right-hand-side of (6.2), coincides with the value random field $\Phi(\underset{\sim}{s}; M)$ of (4.7). We have preferred here to follow a more direct and "natural" approach, which uses the completeness of "write-off" policies and the resulting optimality of $\tau^*(\underset{\sim}{s}, M)$ in (5.3), to "explain" the remarkable equality (6.1) (rather than to introduce it as an Ansatz, as in Whittle (1980) or El Karoui & Karatzas (1992)).

7. A REPRESENTATION OF THE VALUE

Let us recall the notation $\underline{M}_i(\theta) = \min_{0 \le u \le \theta} M_i(u)$ of (3.3), and introduce the *index random field* $\mathcal{M}(\underset{\sim}{s}) \triangleq \max_{1 \le j \le d} M_j(s_j)$ and its *lower envelope*

$$\underline{\mathcal{M}}(\underset{\sim}{s}) \triangleq \min_{\substack{\underset{\sim}{0} \le \underset{\sim}{r} \le \underset{\sim}{s} \\ \underset{\sim}{r} \epsilon S}} \mathcal{M}(\underset{\sim}{r}) = \max_{1 \le j \le d} \underline{M}_j(s_j), \quad \underset{\sim}{s} \epsilon S \tag{7.1}$$

as in Mandelbaum (1986). For an arbitrary allocation strategy $\underset{\sim}{I} \epsilon \mathcal{A}(\underset{\sim}{0})$ of the index-type, we have the following extensions of (3.6), (3.8) with $t = 0$:

$$\underline{\mathcal{M}}(\underset{\sim}{I}(\theta)) = \inf\{m \ge 0 / \sum_{i=1}^{d} \sigma_i(0; m) \le \theta\}\ , \quad \theta \epsilon \mathbf{N}_0 \tag{7.2}$$

$$\begin{aligned} \Phi(\underset{\sim}{0}) &\equiv E \sum_{\theta=0}^{\infty} \beta^\theta \sum_{i=1}^{d} h_i(I_i(\theta+1))[I_i(\theta+1) - I_i(\theta)] \\ &= (1-\beta) \cdot E \sum_{\theta=0}^{\infty} \beta^\theta \underline{\mathcal{M}}(\underset{\sim}{I}(\theta))\ . \end{aligned} \tag{7.3}$$

The representation (7.3) for the value $\Phi(\underset{\sim}{0})$ of the dynamic allocation problem (4.4) with $\underset{\sim}{s} = \underset{\sim}{0}$ is due to Mandelbaum (1986), who derived it using different methodology. As was noticed in that paper, (7.3) also exhibits $\underset{\sim}{I}$ as the solution to a *dual minimization problem,* in the following manner:

7.1 Corollary: *With* $\check{\mathcal{R}}(\underset{\sim}{T}) \triangleq (1-\beta)\sum_{\theta=0}^{\infty} \beta^{\theta} \underline{\mathcal{M}}(\underset{\sim}{T}(\theta)),$ *we have*

$$ E\mathcal{R}(\underset{\sim}{T}') \leq E\mathcal{R}(\underset{\sim}{I}') = \Phi(\underset{\sim}{0}) = E\check{\mathcal{R}}(\underset{\sim}{I}) \leq E\check{\mathcal{R}}(\underset{\sim}{T}) \tag{7.4} $$

where $\underset{\sim}{T}, \underset{\sim}{T}'$ *are arbitrary, and* $\underset{\sim}{I}, \underset{\sim}{I}'$ *are index-type, allocation strategies in* $\mathcal{A}(\underset{\sim}{0})$.

It would be very interesting to have an independent, *direct* proof of the comparison $E\mathcal{R}(\underset{\sim}{T}') \leq E\check{\mathcal{R}}(\underset{\sim}{T})$ in (7.4). As noted by Mandelbaum (1986), this would lead to an alternative proof for the optimality of index-type strategies, in conjunction with the "easy" comparison $E\check{\mathcal{R}}(\underset{\sim}{I}) \leq E\check{\mathcal{R}}(\underset{\sim}{T})$ in (7.4). This still seems to be an open question.

8. PROOFS

Proof of Lemma 2.1: The decrease follows readily from the decrease of the convex function $m \mapsto \varphi(t;m)$, given by

(8.1)

$$ \begin{aligned} \varphi(t;m) &\triangleq (V(t;m)-m)\beta^t \\ &= \operatorname{esssup}_{\tau \in \mathcal{S}(t)} E\Big[\sum_{u=t}^{\tau-1} \beta^u \{h(u+1) - m(1-\beta)\} \,\Big|\, \mathcal{F}(t)\Big] . \end{aligned} $$

Hence, $\sigma_* \triangleq \lim_{\delta \downarrow 0} \sigma_t(m+\delta) \leq \sigma_t(m)$, a.s. On the other hand, we have: $0 \leq \varphi(\sigma_t(m+\frac{1}{k}); m+\delta) \leq \varphi(\sigma_t(m+\frac{1}{k}); m+\frac{1}{k}) = 0$ for $k > 1/\delta, k \epsilon \mathbf{N}$. Letting $k \nearrow \infty$ we get $\varphi(\sigma_*; m+\delta) = 0, \forall \delta > 0$, and from the continuity of $m \mapsto \varphi(t;m)$ we obtain $\varphi(\sigma_*;m) = 0$, whence $\sigma_t(m) \leq \sigma_*$, almost surely.

Proof of Lemma 2.2: Only (2.9) needs discussion, so we fix $m \epsilon [0,\infty)$, recall

$$ V(t;m) = E\Big[m\beta^{\sigma_t(m)-t} + \sum_{u=t}^{\sigma_t(m)-1} \beta^{u-t} h(u+1) \,\Big|\, \mathcal{F}(t)\Big] , \tag{8.2} $$

and observe that for $\delta > 0$ we have

$$Z(t;m) = E[Z(\sigma_t(m+\delta);m)|\mathcal{F}(t)] \,, \quad a.s. \tag{8.3}$$

from (2.6), Lemma 2.1 and optional sampling. This last equation yields, in conjunction with (8.2) (with m replaced by $m+\delta$):
(8.4)

$$\begin{aligned} V(t;m) &= E\Big[\beta^{\sigma_t(m+\delta)-t}\; V(\sigma_t(m+\delta);m) \;+\; \sum_{u=t}^{\sigma_t(m+\delta)-1} \beta^{u-t}h(u+1)|\mathcal{F}(t)\Big] \\ &\geq E\Big[m\beta^{\sigma_t(m+\delta)-t} + V(t;m+\delta) - (m+\delta)\beta^{\sigma_t(m+\delta)-t}|\mathcal{F}(t)\Big] \,. \end{aligned}$$

Thanks to Lemma 2.1, we obtain from this

$$\overline{\lim}_{\delta\downarrow 0}\frac{V(t;m+\delta)-V(t;m)}{\delta} \leq E\Big[\beta^{\sigma_t(m)-t}\;\Big|\;\mathcal{F}(t)\Big] \,, \quad a.s. \tag{8.5}$$

On the other hand, the supermartingale property of $Z(\cdot;m+\delta)$ gives $Z(t;m+\delta) \geq E[Z(\sigma_t(m);m+\delta)|\mathcal{F}(t)]$, a.s. Using this and (8.2), we obtain

$$\begin{aligned} V(t;m+\delta) &\geq E\Big[V(\sigma_t(m);m+\delta)\beta^{\sigma_t(m)-t} + \sum_{u=t}^{\sigma_t(m)-1} \beta^{u-t}h(u+1)\;\Big|\;\mathcal{F}(t)\Big] \\ &\geq E\Big[(m+\delta)\beta^{\sigma_t(m)-t} + V(t;m) - m\beta^{\sigma_t(m)-t}\;\Big|\;\mathcal{F}(t)\Big] \end{aligned}$$

almost surely, whence

$$\underline{\lim}_{\delta\downarrow 0}\frac{V(t;m+\delta)-V(t;m)}{\delta} \geq E\Big[\beta^{\sigma_t(m)-t}\;\Big|\;\mathcal{F}(t)\Big] \,, \quad a.s. \tag{8.6}$$

The conclusion follows from (8.5), (8.6).

Proof of the "forwards induction" characterization (3.2): For any $\tau\epsilon\mathcal{S}(t+1)$ we have almost surely

$$M(t) = V(t;M(t)) \geq E\Big[\sum_{u=t}^{\tau-1}\beta^{u-t}h(u+1) + M(t)\beta^{\tau-t}\;\Big|\;\mathcal{F}(t)\Big] \,,$$

whence $(1-\beta)M(t) \geq \text{esssup}_{\tau\epsilon\mathcal{S}(t+1)} W(t,\tau)$ with

$$W(t,\tau) \triangleq \frac{E\left[\sum_{u=t}^{\tau-1} \beta^u h(u+1) \,\middle|\, \mathcal{F}(t)\right]}{E\left[\sum_{u=t}^{\tau-1} \beta^u \,\middle|\, \mathcal{F}(t)\right]}.$$

On the other hand, with $\xi_n \triangleq M(t) - \frac{1}{n}$ we have $\sigma_t(\xi_n)\epsilon\mathcal{S}(t+1)$ and $\sigma_t(\xi_n) \equiv \sigma_t^*(\xi_n)$, $\forall\, n\epsilon\mathbf{N}$ sufficiently large, amost surely, where $\sigma_t^*(m) := \inf\{\theta \geq t+1/M(\theta) \leq m\}$. Therefore,

$$\xi_n < V(t;\xi_n) = E\Big[\sum_{u=t}^{\tau-1} \beta^{u-t} h(u+1) + \xi_n \beta^{\tau-t} \,\Big|\, \mathcal{F}(t)\Big] \Big|_{\tau=\sigma_t^*(\xi_n)},$$

whence $(1-\beta)\xi_n < W(t,\sigma_t^*(\xi_n))$, $\forall\, n\epsilon\mathbf{N}$. Letting $n \to \infty$ we obtain

$$(1-\beta)\cdot M(t) \leq \text{esssup}_{\tau\epsilon\mathcal{S}(t+1)} W(t,\tau)\ ,\ a.s.$$

This establishes (3.2).

Proof of (3.7) - (3.9): From (3.4) we obtain

$$\beta^{\sigma_t(m)-t} = (1-\beta)\sum_{k=0}^{\infty} \beta^k 1_{\{\sigma_t(m)\leq k+t\}} = (1-\beta)\sum_{\theta=t}^{\infty} \beta^{\theta-t} 1_{\{m\geq \underline{M}(t,\theta)\}},$$

and thus from (2.9), Fubini's theorem, and $V(t;K) = K$:

$$\text{(8.7)}\quad K - V(t;m) = (1-\beta)\cdot E\Big[\sum_{\theta=t}^{\infty} \beta^{\theta-t}(K - (m \vee \underline{M}(t,\theta))) \,\Big|\, \mathcal{F}(t)\Big],$$

and we get (3.9) for $0 \leq m \leq K$; on the other hand, both sides of (3.9) are equal to m, for $m \geq K$. Now set $m = 0$ in (3.9) and recall $\sigma_t(0) = \infty$, to obtain (3.8). Again from (3.9), we have, almost surely:

$$\begin{aligned} V(t;m) - V(t;0) &= (1-\beta)\cdot E\Big[\sum_{\theta=t}^{\infty} \beta^{\theta-t}(m - \underline{M}(t,\theta))^+ \,\Big|\, \mathcal{F}(t)\Big] \\ &= (1-\beta)\cdot E\Big[\sum_{\theta=\sigma_t(m)}^{\infty} \beta^{\theta-t}(m - \underline{M}(t,\theta)) \,\Big|\, \mathcal{F}(t)\Big]. \end{aligned}$$

On the other hand, we obtain from (8.2) and $\sigma_t(0) = \infty$:

$$V(t;m) - V(t;0) = E\Big[m\beta^{\sigma_t(m)-t} - \sum_{\theta=\sigma_t(m)}^{\infty} \beta^{\theta-t}h(\theta+1)\,\Big|\,\mathcal{F}(t)\Big]\ , a.s.$$

Now (3.7) follows by comparing these two expressions.

Proof of Lemma 5.1 *(inspired by Menaldi & Robin (1990)):* The first inequality is obvious; to prove the second, it suffices to show that the **F**-adapted random field $W(\underset{\sim}{s}) \stackrel{\Delta}{=} \Psi^{(i)}(\underset{\sim}{s}^{(i)};M)+V_i(s_i;M)-M\ , \underset{\sim}{s}\epsilon S$ satisfies

$$\begin{aligned} &W(\underset{\sim}{s}) \geq M \\ (8.8)\quad &W(\underset{\sim}{s}) \geq \max_{1\leq j\leq d}\{h_j(s_j+1) + \beta\cdot E[W(\underset{\sim}{s}+\underset{\sim}{e}_j)|\mathcal{F}(\underset{\sim}{s})]\}\ , \quad \forall\ \underset{\sim}{s}\epsilon S \end{aligned}$$

almost surely; because by (4.9), $\Phi(\cdot\ ;M)$ is the smallest **F**-adapted random field that satisfies the inequalities (8.8).
Now $\Psi^{(i)}(\cdot\ ;M) \geq M, V_i(\cdot\ ;M) \geq M$ and so the first inequality in (8.8) is obvious. For the second, take into account the independence of the filtrations $\mathbf{F}_i, \mathbf{F}^{(i)}$ and observe that we have

$$\begin{aligned} &h_i(s_i+1) + \beta\cdot E[W(\underset{\sim}{s}+\underset{\sim}{e}_i)|\mathcal{F}(\underset{\sim}{s})] \\ &\quad = h_i(s_i+1) + \beta\cdot E[W(\underset{\sim}{s}+\underset{\sim}{e}_i)|\mathcal{F}_i(s_i)\vee\mathcal{F}^{(i)}(\underset{\sim}{s}^{(i)})] \\ &\quad = h_i(s_i+1) + \beta\cdot E[V_i(s_i+1;M)|\mathcal{F}_i(s_i)] + \beta[\Psi^{(i)}(\underset{\sim}{s}^{(i)};M) - M] \\ &\quad \leq V_i(s_i;M) + \Psi^{(i)}(\underset{\sim}{s}^{(i)};M) - M = W(\underset{\sim}{s})\ , \quad \forall\ \underset{\sim}{s}\epsilon S\ \ a.s. \end{aligned}$$

from (2.7), whilst for any $j \neq i$:

$$\begin{aligned} &h_j(s_j+1) + \beta\cdot E[W(\underset{\sim}{s}+\underset{\sim}{e}_j)|\mathcal{F}(\underset{\sim}{s})] \\ &\quad = h_j(s_j+1) + \beta\cdot E[W(\underset{\sim}{s}+\underset{\sim}{e}_j)|\mathcal{F}_i(s_i)\vee\mathcal{F}^{(i)}(\underset{\sim}{s}^{(i)})] \\ &\quad = h_j(s_j+1) + \beta\cdot E[\Psi^{(i)}(\underset{\sim}{s}^{(i)}+\underset{\sim}{e}_j;M)|\mathcal{F}^{(i)}(\underset{\sim}{s}^{(i)})] + \beta[V_i(s_i;M) - M] \\ &\quad \leq \Psi^{(i)}(\underset{\sim}{s}^{(i)};M) + V_i(s_i;M) - M = W(\underset{\sim}{s})\ , \quad \forall\ \underset{\sim}{s}\epsilon S\ \ a.s. \end{aligned}$$

from (5.1). This proves the second inequality in (8.8).

Proof of Proposition 5.2: Take an arbitrary $\Pi = (\underset{\sim}{T},\tau)\ \epsilon\ \mathcal{P}(\underset{\sim}{s})$.
(a) Suppose that for some $t\epsilon\mathbf{N}_0$, we have $V_i(T_i(t);M) = M$ for some $i\epsilon\{1,\ldots,d\}$. Then Lemma 5.1 gives $\Phi(\underset{\sim}{T}(t);M) = \Psi^{(i)}(\underset{\sim}{T}^{(i)}(t);M)$;

in other words, *at time t it is then optimal to abandon the ith project.*

(b) Suppose that at some $t \epsilon \mathbf{N}_0$, we have $V_i(T_i(t); M) = M$, $\forall\ i = 1, \ldots, d$. Then, as before, $\Phi(\underset{\sim}{T}(t); M) = \Psi^{(i)}(\underset{\sim}{T}^{(i)}(t); M)$, $\forall\ i = 1, \ldots, d$, and continuing this way we conclude $\Phi(\underset{\sim}{T}(t); M) = M$. In other words, *it is optimal then to retire at time t.*

(c) Finally, suppose that at some $t \epsilon \mathbf{N}_0$, we have $V_i(T_i(t); M) > M$ for some $i \epsilon \{1, \ldots, d\}$. Then from (5.2) we get, in particular, $\Phi(\underset{\sim}{T}(t); M) \geq V_i(T_i(t); M) > M$; in other words, *it is not optimal to retire as long as there is at least one project that has not been abandoned.*

Putting these three conclusions together, we obtain (5.4).

Proof of Proposition 6.1: Only (6.1) need be discussed. For any given $\Pi = (\underset{\sim}{T}, \tau)$ in the class $\mathcal{P}(\underset{\sim}{s}, M)$, we have that τ is equal to $\tau^*(\underset{\sim}{s}, M)$, the $\{\mathcal{F}(\underset{\sim}{T}(t))\}$-stopping time of (5.3), and therefore

$$
\begin{aligned}
E[\beta^\tau | \mathcal{F}(\underset{\sim}{s})] &= E\Big[\prod_{i=1}^{d} \beta^{\sigma_i(s_i;M)-s_i} \mid \mathcal{F}(\underset{\sim}{s})\Big] \\
&= \prod_{i=1}^{d} E\Big[\beta^{\sigma_i(s_i;M)-s_i} \mid \mathcal{F}_i(s_i)\Big] = \prod_{i=1}^{d} \Big(\frac{\partial^+}{\partial m} V_i(s_i; M)\Big)\ , \quad a.s.
\end{aligned} \tag{8.9}
$$

Thus, taking conditional expectations with respect to $\mathcal{F}(\underset{\sim}{s})$ in $\mathcal{R}(\Pi; M \pm \delta) - \mathcal{R}(\Pi; M) = \pm\delta\beta^\tau$ with $\delta \epsilon (0, M)$, we obtain

$$
\begin{aligned}
\Phi(\underset{\sim}{s}; M \pm \delta) &\geq E[\mathcal{R}(\Pi; M \pm \delta) | \mathcal{F}(\underset{\sim}{s})] \\
&= E[\mathcal{R}(\Pi; M) | \mathcal{F}(\underset{\sim}{s})] \pm \delta . \prod_{i=1}^{d} \Big(\frac{\partial^+}{\partial m} V_i(s_i; M)\Big)\ ,
\end{aligned}
$$

and taking the supremum of the right-hand side over $\Pi \epsilon \mathcal{P}^*(\underset{\sim}{s}, M)$,

$$
\Phi(\underset{\sim}{s}; M \pm \delta) \geq \Phi(\underset{\sim}{s}; M) \pm \delta \cdot \prod_{i=1}^{d} \Big(\frac{\partial^+}{\partial m} V_i(s_i; M)\Big) \tag{8.10}
$$

thanks to (5.5). This leads directly to

$$\underline{\lim}_{\delta\downarrow 0}\frac{1}{\delta}[\Phi(\underset{\sim}{s};M+\delta)-\Phi(\underset{\sim}{s};M)] \ge \prod_{i=1}^{d}\Big(\frac{\partial^+}{\partial m}V_i(s_i;m)\Big)$$
$$\ge \overline{\lim}_{\delta\downarrow 0}\frac{1}{\delta}[\Phi(\underset{\sim}{s};M)-\Phi(\underset{\sim}{s};M-\delta)]$$

and thence to (6.1), thanks to the convexity of the map $M \mapsto \Phi(\underset{\sim}{s};M)$.

Proof of (6.4): Fix $i\epsilon\{1,\dots,d\}$; for given $\underset{\sim}{s}^{(i)} \triangleq (s_1,\dots,s_{i-1}, s_{i+1},\dots,s_d)$ in $\mathbf{N}_0^{d-1}$, the function $m \mapsto P_i(\underset{\sim}{s}^{(i)};m) \triangleq \prod_{j\neq i}\frac{\partial^+}{\partial m} V_j(s_j;m)$ is increasing, and integrating by parts in (6.2) we obtain

$$\Phi(\underset{\sim}{s};M) = V_i(s_i;M)P_i(\underset{\sim}{s}^{(i)};M)+\int_M^K V_i(s_i;m)d_mP_i(\underset{\sim}{s}^{(i)};m)\,. \tag{8.11}$$

Thus:
$\Phi(\underset{\sim}{s}+\underset{\sim}{e}_i;M) = V_i(s_i+1;M)P_i(\underset{\sim}{s}^{(i)};M)+\int_M^K V_i(s_i+1;m)d_mP_i(\underset{\sim}{s}^{(i)};m)$,

$$E\Big[\Phi(\underset{\sim}{s}+\underset{\sim}{e}_i;M)\;\Big|\;\mathcal{F}(\underset{\sim}{s})\Big] = P_i(\underset{\sim}{s}^{(i)};M)\cdot E\Big[V_i(s_i+1;M)\;\Big|\;\mathcal{F}_i(s_i)\Big]+$$
$$+\int_M^K E\Big[V_i(s_i+1;m)\;\Big|\;\mathcal{F}_i(s_i)\Big]d_mP_i(\underset{\sim}{s}^{(i)};m)\,,$$

whence

$$\begin{aligned}&\Phi(\underset{\sim}{s};M)-h_i(s_i+1)-\beta\cdot E\Big[\Phi(\underset{\sim}{s}+\underset{\sim}{e}_i;M)\;\Big|\;\mathcal{F}(\underset{\sim}{s})\Big] =\\ &= \varphi_i(s_i;M)P_i(\underset{\sim}{s}^{(i)};M)+\int_M^K\varphi_i(s_i;m)d_mP_i(\underset{\sim}{s}^{(i)};m)\,,\end{aligned} \tag{8.12}$$

with $\varphi_i(t;m)\triangleq V_i(t;m)-h_i(t+1)-\beta\cdot E\Big[V_i(t+1;m)\;\Big|\;\mathcal{F}_i(t)\Big]$. From (2.7) this last expression is nonnegative, and thus so is the right-hand side of (8.12), almost surely.

Proof of (6.5): From Remark 3.2 we have $0\le\prod_{i=1}^d\frac{\partial^+}{\partial m}V_i(s_i;m)\le 1$ as well as $\frac{\partial^+}{\partial m}V_i(s_i;m) = 1$ for all $m\ge M, 1\le i\le d$ on the event $\{M\ge\max_{1\le j\le d}M_j(s_j)\}$, and so we have $\Phi(\underset{\sim}{s};M)=M$ on this event.

Proof of (6.6): Consider arbitrary $m\epsilon[M,K]$. On $\{m < M_i(s_i)\}$ we have $\varphi_i(s_i;m) = 0$; whereas on $\{m \geq M_i(s_i)\}$ we have from Remark 3.2, for every $j \neq i$: $m \geq M_j(s_j) \Leftrightarrow \frac{\partial^+}{\partial m}V_j(s_j;m) = 1$, so that $P_i(\underset{\sim}{s}^{(i)};m) = 1$ Thus the right-hand side of (8.12) is equal to zero on $\{M < M_i(s_i) = \max_{1\leq j \leq d} M_j(s_j)\}$.

Proof of Corollaries 6.3, 6.4: For any allocation strategy $\underset{\sim}{T}\epsilon\mathcal{A}(\underset{\sim}{0})$, the random sequence

$$Q(t;\underset{\sim}{T}) \triangleq \beta^t \Phi(\underset{\sim}{T}(t)) + \sum_{u=0}^{t-1} \beta^u \sum_{i=1}^{d} h_i(T_i(u+1))[T_i(u+1) - T_i(u)]$$

is a bounded, nonnegative, $\{\mathcal{G}(t)\}_{t\epsilon\mathbf{N}_0}$ - supermartingale, and it is now easy to see that $Q(\cdot\,;\underset{\sim}{T})$ is a martingale if $\underset{\sim}{T}$ is of index-type. All claims follow readily.

Proof of (7.2), (7.3): For every $\theta\epsilon\mathbf{N}_0$ we have

$$\text{(8.13)} \qquad \sum_{j=1}^{d} \sigma_j(0;m) \leq \theta \iff m \geq \underline{\mathcal{M}}(\underset{\sim}{I}(\theta)) = \max_{1\leq j\leq d} \underline{M}_j(I_j(\theta)),$$

as both statements express that "the index-type policy $\underset{\sim}{I}(\cdot)$ has abandoned all projects $j = 1,\ldots,d$ by time θ". Now (7.2) is just a restatement of (8.13); as for (7.3), notice that (8.13) gives

$$\text{(8.14)} \qquad \begin{aligned} \beta^{\sum_{i=1}^{d}\sigma_i(0;m)} &= (1-\beta)\sum_{\theta=0}^{\infty}\beta^\theta 1_{\{\sum_{i=1}^{d}\sigma_i(0;m)\leq\theta\}} \\ &= (1-\beta)\sum_{\theta=0}^{\infty}\beta^\theta 1_{\{m\geq\underline{\mathcal{M}}(\underset{\sim}{I}(\theta))\}}\,. \end{aligned}$$

Integrating over $[0,K]$ and then taking expectations, we obtain

$$\text{(8.15)} \qquad \begin{aligned} E\int_0^K \beta^{\sum_{i=1}^{d}\sigma_i(0;m)}dm &= (1-\beta)\cdot E\sum_{\theta=0}^{\infty}\beta^\theta[K - \underline{\mathcal{M}}(\underset{\sim}{I}(\theta))] \\ &= K - (1-\beta)\cdot E\sum_{\theta=0}^{\infty}\beta^\theta\underline{\mathcal{M}}(\underset{\sim}{I}(\theta)) \end{aligned}$$

and (7.3) follows from this, in conjuction with (6.2) and (2.9).

Proof of Corollary 7.1: The equalities and the first inequality in (7.4) are obvious, from (7.3) and Corollary 6.4, respectively. It remains to prove the second inequality.

For an arbitrary strategy $\underset{\sim}{T}(\cdot)\epsilon\mathcal{A}(\underset{\sim}{0})$, let $T_i^{-1}(u) \triangleq \min\{t\epsilon\mathbf{N}_0 \ / \ T_i(t) \geq u\}$ and observe that we have from (7.1), (3.4), and with $\sigma_j(m) \equiv \sigma_j(0;m)$:

$$m \geq \underset{\sim}{\mathcal{M}}(\underset{\sim}{T}(\theta)) \Leftrightarrow \sigma_j(m) \leq T_j(\theta) \ , \qquad \forall\, j = 1, \ldots, d$$
$$\Leftrightarrow \theta \geq \max_{1\leq j \leq d} T_j^{-1}(\sigma_j(m)) =: \Theta(m; \underset{\sim}{T}) \ .$$

Therefore, we obtain by analogy with (8.14), (8.15):

$$\frac{\beta^{\Theta(m;\underset{\sim}{T})}}{1-\beta} = \sum_{\theta=0}^{\infty} \beta^{\theta}\, 1_{\{m \geq \underset{\sim}{\mathcal{M}}(\underset{\sim}{T}(\theta))\}} \text{ and } \int_0^K \beta^{\Theta(m;\underset{\sim}{T})} dm = K - \check{\mathcal{R}}(\underset{\sim}{T}) \ ,$$

almost surely. In order to prove (7.4) it suffices then to show that $\Theta(m; \underset{\sim}{T}) \geq \sum_{j=1}^d \sigma_j(m)$ holds *on any given path* $\omega\epsilon\Omega$.

Suppose the contrary; i.e., that on the chosen $\omega\epsilon\Omega$ we have $T_i^{-1}(\sigma_i(m)) < \sum_{j=1}^d \sigma_j(m)$, $\forall\, i \in \{1, \ldots, d\}$. This implies then $\sigma_i(m) \leq T_i(\sum_{j=1}^d \sigma_j(m))$, $\forall\, i\epsilon\{1, \ldots, d\}$, with the inequality strict for at least one such integer i; whence $\sum_{i=1}^d \sigma_i(m) < \sum_{i=1}^d T_i(\sum_{j=1}^d \sigma_j(m)) = \sum_{j=1}^d \sigma_j(m)$, a contradiction.

REFERENCES

CHOW, Y.S., ROBBINS, H. & SIEGMUND, D. (1991) *The Theory of Optimal Stopping.* Second Edition, Dover, New York.

EL KAROUI, N. & KARATZAS, I. (1992) General Gittins index processes in discrete time. *Proceedings of the National Academy of Sciences (USA)*, to appear.

GITTINS, J.C. (1979) Bandit processes and dynamic allocation indices (with discussion). *J. Roy. Statist. Soc.* **B 41**, 148-164.

GITTINS, J.C. (1989) *Multi-Armed Bandits and Allocation Indices.* J. Wiley & Sons, Chichester.

MANDELBAUM, A. (1986) Discrete multi-armed bandits and multi-parameter processes. *Probab. Theory Rel. Fields* **71**, 129-147.

MANDELBAUM, A. & VANDERBEI, R.J. (1981) Optimal stopping and semimartingales over partially-ordered sets. *Z. Wahrscheinlichkeitstheorie verw. Gebiete* **57**, 253-264.

MENALDI, J.L. & ROBIN, M. (1990) On the optimal reward function of the continuous-time, multi-armed bandit problem. *SIAM J. Control & Optimization* **28**, 97-112.

NEVEU, J. (1975) *Discrete-parameter Martingales.* English Translation, North-Holland, Amsterdam.

TSITSIKLIS, J. (1986) A lemma on the multi-armed bandit problem. *IEEE Trans. Autom. Control* **AC-31**, 576-577.

VARAIYA, P., WALRAND, J. & BUYUKKOC, C. (1985) Extensions of the multi-armed bandit problem: the discounted case. *IEEE Trans. Autom. Control* **AC-30**, 426-439.

WHITTLE, P. (1980) Multi-armed bandits and the Gittins index. *J. Roy. Statist. Soc.* **B 42**, 143-149.

WHITTLE, P. (1982) *Optimization over Time: Dynamic Programming and Stochastic Control.* Wiley, New York.

ON THE STRONG MARKOV PROPERTY OF ONE-DIMENSIONAL CONTINUOUS MARKOV PROCESSES

H.J. ENGELBERT
Friedrich Schiller University, Faculty of Mathematics, Institute for Stochastics, D-O-6900 Jena, Germany

Abstract The explicit form of the fine topology of one-dimensional continuous MARKOV processes is given. On this basis, necessary and sufficient conditions for the strong MARKOV property are stated. It turns out that the fine topology as well as the necessary and sufficient conditions for the strong MARKOV property only depend on the finite dimensional distribution or, in other words, on the semigroup of transition probabilities of the process.

Key Words Continuous MARKOV processes, strong MARKOV property, fine topology, standard semigroups, right processes, perfect processes

1 INTRODUCTION

Let R be the real line, $\mathcal{B}(R)$ the σ-algebra of Borel subsets of R, and $\mathcal{B}^u(R)$ its universal completion.

We consider a stochastic process X which is defined on a family $(\Omega, \mathcal{F}, \mathbf{P}_x, x \in R)$ of probability spaces and adapted to a filtration $\mathbb{F} = (\mathcal{F}_t)_{t \geq 0}$ of sub-σ-algebras of $\mathcal{F}$, with values in the state space $(R, \mathcal{B}^u(R))$.

We will assume that $(X, \mathbb{F})$ is a (homogeneous and conservative)

MARKOV process in the state space $(R, \mathcal{B}(R))$ (resp. $(R, \mathcal{B}^u(R))$), i.e., it satisfies the following conditions:

(i) $\mathbf{P}_x(X_0 = x) = 1$ for all $x \in R$.

(ii) For all $t \geq 0$ and $A \in \mathcal{B}(R)$, $\mathbf{P}_x(X_t \in A)$ is a $\mathcal{B}(R)$-measurable (resp. $\mathcal{B}^u(R)$-measurable) function of x.

(iii) For all $s, t \geq 0, x \in R$, and $A \in \mathcal{B}(R)$

$$\mathbf{P}_x(X_{s+t} \in A \mid \mathcal{F}_s) = \mathbf{P}_{X_s}(X_t \in A) \qquad \mathbf{P}_x\text{-a.s.}$$

In the following, for all unexplained notions, see the standard literature on MARKOV processes, e.g., R.M. Blumenthal & R.K. Getoor [1], E.B. Dynkin [2], H.J. Engelbert [3], R.K. Getoor [5], M.J. Sharpe [9].

Let us consider the σ-algebras $\mathcal{G}^0$ and $\mathcal{G}_t^0$ generated by $\{X_s, s \geq 0\}$ and $\{X_s, s \leq t\}$, respectively. The σ-algebras $\mathcal{G}$ and $\mathcal{G}_t$ are introduced by completing $\mathcal{G}^0$ and $\mathcal{G}_t^0$ in the usual way (cf. [3], I.13).

In the present note, unless otherwise stated we always assume that X possesses continuous paths.

Let $\mathbf{B}$ (resp. $\mathbf{B}^u$) be the space of bounded measurable functions defined on $(R, \mathcal{B}(R))$ (resp. $(R, \mathcal{B}^u(R))$). For every $f \in \mathbf{B}^u, x \in R$, and $t \geq 0$ we set

$$P_t(x, f) = \mathbf{E}_x f(X_t)$$

and denote the function $x \to P_t(x, f)$ by $P_t f$. Then the family $(P_t)_{t \geq 0}$ is a semigroup of positive linear operators acting in $\mathbf{B}$ (resp. $\mathbf{B}^u$) and continuous with respect to the monotone convergence. The family $(P_t)_{t \geq 0}$ is called the semigroup of transition probabilities of $(X, \mathbb{F})$.

Note that the semigroup $(P_t)_{t \geq 0}$ of $(X, \mathbb{F})$ is *natural* in the sense of [3], Definition II.1. (Recall that a semigroup $(P_t)_{t \geq 0}$ is said to be natural if for the space $\mathbf{B}_0$ (resp. $\mathbf{B}_0^u$) of all $f \in \mathbf{B}$ (resp. $\mathbf{B}^u$) such that $\lim_{t \downarrow 0} P_t(x, f) = f(x)$ for every $x \in R$ the relation $\mathcal{B}(R) = \sigma(\mathbf{B}_0)$ (resp. $\mathcal{B}(R) \subseteq \sigma(\mathbf{B}_0^u)$) holds.) Indeed, because of the continuity of the paths of X, every bounded continuous function belongs to $\mathbf{B}_0 \subseteq \mathbf{B}_0^u$.

We will need the notion of a *standard semigroup* which was introduced by the author in [3] (Definition II.3).

Let $\mathbf{C}_s$ (resp. $\mathbf{C}_s^u$) be the space of all bounded $\mathcal{B}(R)$-measurable (resp. $\mathcal{B}^u(R)$-measurable) functions that are continuous in the topology $\mathcal{C}_s$ (resp. $\mathcal{C}_s^u$) of stochastic continuity (see [3], Definition I.18).

By Theorem I.27 from [3], a bounded $(R, \mathcal{B}(R))$-measurable (resp. $(R, \mathcal{B}^u(R))$-measurable) function f belongs to $\mathbf{C}_s$ (resp. $\mathbf{C}_s^u$) if and only if the process $f \circ X$ is $\mathbf{P}_x$-stochastic continuous at the right for every $x \in R$.

DEFINITION 1 A semigroup $(P_t)_{t>0}$ is called a *standard semigroup* in $(R, \mathcal{B}(R))$ (resp. in $(R, \mathcal{B}^u(R))$) if the following conditions are satisfied:

(i) $\mathcal{B}(R) = \sigma(\mathbf{C}_s)$ (resp. $\mathcal{B}(R) \subseteq \sigma(\mathbf{C}_s^u)$).

(ii) $P_t\, \mathbf{C}_s \subseteq \mathbf{C}_s$ (resp. $P_t\, \mathbf{C}_s^u \subseteq \mathbf{C}_s^u$), $t \geq 0$.

We will often use the hypothesis that the semigroup $(P_t)_{t \geq 0}$ of $(X, \mathbb{F})$ is a standard semigroup. It can easily be verified that then the semigroup $(P_t)_{t \geq 0}$ is *measurable*, i.e., for all bounded $\mathcal{B}(R)$-measurable functions f the function $(t, x) \to P_t(x, f)$ is $\mathcal{B}(R_+) \times \mathcal{B}(R)$-measurable (resp. $\mathcal{B}(R_+) \times \mathcal{B}^u(R)$-measurable).* Note that the condition (i) is only stated for completeness. In the present situation it is always satisfied since, because of the continuity of the paths of $(X, \mathbb{F})$, every bounded continuous function belongs to $\mathbf{C}_s$.

Equivalent conditions to the hypothesis that $(P_t)_{t \geq 0}$ is a standard semigroup will be given below. For the proof, cf. [3], Theorem II.7.

PROPOSITION 1 *Let $(X, \mathbb{F})$ be a continuous* MARKOV *process with values in $(R, \mathcal{B}(R))$ (resp. $(R, \mathcal{B}^u(R))$) and $(P_t)_{t \geq 0}$ its semigroup of transition probabilities. The following conditions are equivalent:*

(i) *$(P_t)_{t \geq 0}$ is a standard semigroup in $(R, \mathcal{B}(R))$ (resp. $(R, \mathcal{B}^u(R))$).*

(ii) *$(X, \mathbb{F}_+)$ is a* MARKOV *process in $(R, \mathcal{B}(R))$ (resp. $(R, \mathcal{B}^u(R))$).*†

(iii) *For every $t \geq 0$, $\mathcal{G}_t = \mathcal{G}_{t+}$.*

(iv) $\mathcal{G}_0 = \mathcal{G}_{0+}$.

(v) *Every $\mathcal{B}(R)$-measurable (resp. $\mathcal{B}^u(R)$-measurable) r-excessive function is C_s-continuous (resp. C_s^u-continuous).*

(vi) *For every r-excessive function f the function $f \wedge 1$ is r-excessive, too.*

---

*For a topological space E, $\mathcal{B}(E)$ denotes the σ-algebra of BOREL subsets of E.

†For any filtration $\mathbb{H}$, by $\mathbb{H}_+$ we denote the smallest right continuous filtration containing $\mathbb{H}$.

(vii) *For all $\mathcal{B}(R)$-measurable (resp. $\mathcal{B}^u(R)$-measurable) r-excessive functions f and g the function $f \wedge g$ is again r-excessive.*

COROLLARY 1 *If $(X, \mathbb{F})$ admits a standard semigroup in $(R, \mathcal{B}(R))$ (resp. $(R, \mathcal{B}^u(R))$) then the 0–1 law of* BLUMENTHAL *is satisfied: The σ-algebra $\mathcal{G}_{0+}$ is $\mathbf{P}_x$-trivial.*

BLUMENTHAL's 0–1 law is the crucial point for the classification of the states of $(X, \mathbb{F})$ which can now be realized as in K. Itô & H.P. McKean [6].

For the remainder of this section we assume that $(X, \mathbb{F})$ admits a standard semigroup.

For all $a \in R$, let us define the following random times:

$$U(a+) = \inf\{t \geq 0 : X_t > a\},$$

$$U(a-) = \inf\{t \geq 0 : X_t < a\}.$$

It is easy to see that $U(a+)$ and $U(a-)$ are $\mathbb{G}_+$-stopping times where $\mathbb{G}$ is the filtration $(\mathcal{G}_t)_{t \geq 0}$. Hence

$$p_+(a) = \mathbf{P}_a(U(a+) = 0), \qquad p_-(a) = \mathbf{P}_a(U(a-) = 0)$$

can only take the values 0 or 1.

DEFINITION 2 (i) A state $x \in R$ is called *regular* if the equality $p_+(x) = p_-(x) = 1$ holds.

(ii) A state $x \in R$ is said to be *left singular* (resp. *right singular*) if $p_+(x) = 0$ (resp. $p_-(x) = 0$).

(iii) A state $x \in R$ is called a *holding point* if $p_+(x) = p_-(x) = 0$.

By K_r, K_-, K_+, H we denote, respectively, the set of regular, left singular, right singular, and holding points. Obviously,

$$R = K_r \cup K_- \cup K_+, \qquad K_r \cap (K_- \cup K_+) = \emptyset, \qquad H = K_- \cap K_+.$$

A state $x \in K_s := K_- \cup K_+$ is called *singular* .

REMARK 1 The classification of the states given in Definition 2 can also be realized if we replace the condition that $(X, \mathbb{F})$ admits a standard semigroup by the following hypothesis: *For every $x \in R$, $\mathbf{P}_x(U(x+) > 0)$ and $\mathbf{P}_x(U(x-) > 0)$ only take the values* 0 *or* 1. This will be used several times in the sequel.

2 THE FINE TOPOLOGY

The aim of the present section is to describe the fine topology $\mathcal{C}_0$ associated with the continuous MARKOV process $(X, \mathbb{F})$. To this end we assume that the semigroup $(P_t)_{t\geq 0}$ of $(X, \mathbb{F})$ is a standard semigroup in $(R, \mathcal{B}^u(R))$.

DEFINITION 3 A subset G of R is called *finely open* (with respect to the state space $(R, \mathcal{B}(R))$) if for all $x \in G$ there exists a set $A \in \mathcal{B}(R)$ such that $x \in A \subseteq G$ and for the first entry time [‡] D_{A^c} of X into A^c [§] the conditions $\{D_{A^c} > 0\} \in \mathcal{F}$ and $\mathbf{P}_x(D_{A^c} > 0) = 1$ hold. By $\mathcal{C}_0$ we denote the collection of all finely open sets. The system $\mathcal{C}_0$ is called the *fine topology of* $(X, \mathbb{F})$ (with respect to the state space $(R, \mathcal{B}(R))$).

THEOREM 1 *Let $G \subseteq R$. Then $G \in \mathcal{C}_0$ if and only if for every $x \in G$ which is not left singular (resp. not right singular) there exists a right (resp. left) neighbourhood U_x of x such that $U_x \subseteq G$.*

COROLLARY 1 (i) *If x is regular then the intervals $(x - r, x + r)$ for rational numbers $r > 0$ form a base for the fine topology $\mathcal{C}_0$ at x.*

(ii) *If x is right singular (resp. left singular) but not left singular (resp. not right singular) then the intervals $[x, x + r)$ (resp. $(x - r, x]$) for rational numbers $r > 0$ form a base for the fine topology $\mathcal{C}_0$ at x.*

(iii) *If x is a holding point then $\{x\} \in \mathcal{C}_0$.*

In view of Corollary 1, the fine topology $\mathcal{C}_0$ satisfies the first axiom of countability but, in general, it does not satisfy the second axiom of countability.

COROLLARY 2 *A sequence (x_n) of real numbers converges to a point $x \in R$ in the fine topology $\mathcal{C}_0$ if and only if:*

(i) $\lim_n x_n = x$ *(in the usual topology).*

(ii) *If $x \in K_+$ (resp. $x \in K_-$) then $x_n \geq x$ (resp. $x_n \leq x$) for all sufficiently large n.*

[‡] For any $B \subseteq R$, $D_B = \inf\{t \geq 0 : X_t \in B\}$.

[§] A^c denotes the complement of the set A.

Corollaries 1 – 2 immediately follow from Theorem 1.

Proof of Theorem 1 Let G be a subset of R satisfying the condition of the theorem. We have to show that G is finely open. For this it is sufficient to verify that for every $x \in G$ there exists a finely open set O_x with $x \in O_x \subseteq G$. Now we fix $x \in G$. If x is not left singular (resp. not right singular), let U_x^+ (resp. U_x^-) be a right (resp. left) neighbourhood of x contained in G, existing by the condition of the theorem. We now set:

$$O_x = \begin{cases} \{x\} & \text{if } x \text{ is a holding point} \\ U_x^+ & \text{if } x \text{ is right singular but not left singular} \\ U_x^- & \text{if } x \text{ is left singular but not right singular} \\ U_x^- \cup U_x^+ & \text{if } x \text{ is regular.} \end{cases}$$

From the right continuity of the paths of X, the definition of right and left singular points, and BLUMENTHAL's 0–1 law (cf. Definition 2 and Corollary 1 to Proposition 1) it is now easy to see that O_x is a fine neighbourhood of x contained in G. This proves the sufficiency of the condition of the theorem.

For proving the necessity of the condition, let G be a finely open subset of R and $x \in G$. Then there exists $A \in \mathcal{B}(R)$ such that $x \in A \subseteq G$ and the event $\{D_{A^c} > 0\}$ has $\mathbf{P}_x$-probability 1. In view of the continuity of X,

$$I(\omega) = \{X_t(\omega) : \; t < D_{A^c}(\omega)\}$$

is an interval. Moreover, $x \in I(\omega)$ and $I(\omega) \subseteq G$ $\mathbf{P}_x$-a.s. If x is not left singular (resp. not right singular) then $I(\omega) \cap (x, +\infty) \neq \emptyset$ (resp. $I(\omega) \cap (-\infty, x) \neq \emptyset$) $\mathbf{P}_x$-a.s. and hence $I(\omega)$ is a right (resp. left) neighbourhood of x contained in G $\mathbf{P}_x$-a.s. This proves the necessity of the condition and hence the theorem. □

REMARKS 2 (i) Analogously we can introduce the fine topology $\mathcal{C}_0^u$ with respect to the state space $(R, \mathcal{B}^u(R))$. Theorem 1 is also true if we replace $\mathcal{C}_0$ by $\mathcal{C}_0^u$. As a consequence we see that $\mathcal{C}_0 = \mathcal{C}_0^u$. This means that the fine topology does not change if we enlarge $\mathcal{B}(R)$ to the σ-algebra $\mathcal{B}^u(R)$ of universally measurable sets.

(ii) Theorem 1 and its corollaries remain true if only the sets $\{U(x+) > 0\}$ and $\{U(x-) > 0\}$ are $\mathbf{P}_x$-trivial for every $x \in R$.

3 RIGHT PROCESSES AND PERFECT PROCESSES

The main purpose of the present note is to ask for conditions guaranteeing the strong MARKOV property of a one-dimensional continuous MARKOV process $(X, \mathbb{F}_+)$.

It turns out that this is equivalent to the property that $(X, \mathbb{F})$ belongs to the class of right MARKOV processes and also to the class of perfect MARKOV processes (cf. H.J. Engelbert [3], R.K. Getoor [5], M.J. Sharpe [9]). This enables us to use general criteria for $(X, \mathbb{F})$ being a right MARKOV process stated in [3], §7.

Let us first recall the notions of right and perfect processes (cf. [3], Definitions IV.1 and V.2).

DEFINITION 4 Let $(X, \mathbb{F})$ be a MARKOV process defined on a family $(\Omega, \mathcal{F}, \mathbf{P}_x, x \in E)$ of probability spaces and taking values in the state space $(E, \mathcal{E})$ (resp. in its universal completion $(E, \mathcal{E}^u)$).

(i) $(X, \mathbb{F})$ is called a right process in $(E, \mathcal{E})$ (resp. $(E, \mathcal{E}^u)$) if the following conditions are satisfied:

  (a) $(X, \mathbb{F})$ has a natural semigroup $(P_t)_{t \geq 0}$ of transition probabilities in $(E, \mathcal{E})$ (resp. $(E, \mathcal{E}^u)$) (cf. Section 1 or [3], Definition II.1).

  (b) For every $r \geq 0$ and $\mathcal{E}$-measurable (resp. $\mathcal{E}^u$-measurable) r-excessive function f the process $f \circ X$ is a.s. right continuous.

(ii) Let $(X, \mathbb{F})$ be a right MARKOV process. Then $(X, \mathbb{F})$ is called a perfect MARKOV process if there exists a process X^- which satisfies the following conditions:

  (a) $X_0^- = X_0$.

  (b) For some $p > 0$ and for all nonnegative bounded $\mathcal{E}$-measurable (resp. $\mathcal{E}^u$-measurable) f we have

$$U_p(X_t^-, f) = \lim_{s \uparrow t} U_p(X_s, f) \text{ for all } t > 0 \quad \text{a.s.}$$

  where

$$U_p(x, f) = \int_0^\infty e^{-pt} P_t(x, f)\, dt$$

  is the p-potential of f at x.

The process X^- is called the *left version* of X. Note that every right MARKOV process $(X, \mathbb{F})$ in $(E, \mathcal{E})$ (resp. $(E, \mathcal{E}^u)$) possesses a standard semigroup $(P_t)_{t\geq 0}$ of transition probabilities in $(E, \mathcal{E})$ (resp. $(E, \mathcal{E}^u)$) (cf. [3], Definition II.3 and Theorem II.7). In the case of an one-dimensional continuous MARKOV process, see Definition 1 and Proposition 1.

It should also be mentioned that the only difference between, for example, right MARKOV processes in $(E, \mathcal{E})$ and right MARKOV processes in $(E, \mathcal{E}^u)$ is that in the first case the corresponding semigroup $(P_t)_{t\geq 0}$ leaves the space $\mathbf{B}$ of bounded measurable functions on $(E, \mathcal{E})$ invariant.

Now we return to our one-dimensional continuous MARKOV process $(X, \mathbb{F})$ in $(R, \mathcal{B}(R))$ (resp. $(R, \mathcal{B}^u(R))$). In the next theorem we do not assume that the semigroup $(P_t)_{t\geq 0}$ of $(X, \mathbb{F})$ is standard.

Let $\mathbb{H} = (\mathcal{H}_t)_{t\geq 0}$ be a filtration contained in $\mathbb{F}_+$. We say that $(X, \mathbb{H})$ possesses the strong Markov property if X is $\mathbb{H}$-adapted and for every $\mathbb{H}$-stopping time T, bounded $\mathcal{B}^u(R)$-measurable function f, $t > 0$ and $x \in R$ we have

$$\mathbf{E}_x\left(f(X_{T+t}) \mid \mathcal{H}_T\right) = P_t(X_T, f) \quad \text{on } \{T < \infty\} \quad \mathbf{P}_x\text{-a.s.} \qquad (1)$$

THEOREM 2 *The following conditions are equivalent:*

(i) (a) $(X, \mathbb{G})$ *possesses the strong* MARKOV *property.*

(b) $(X, \mathbb{F})$ *has a standard semigroup* $(P_t)_{t\geq 0}$ *in* $(R, \mathcal{B}^u(R))$.

(ii) $(X, \mathbb{G}_+)$ *possesses the strong* MARKOV *property.*

(iii) $(X, \mathbb{F})$ *is a right* MARKOV *process in the state space* $(R, \mathcal{B}(R))$.

(iv) $(X, \mathbb{F})$ *is a perfect* MARKOV *process in the state space* $(R, \mathcal{B}(R))$ *with left version* $X^- = X$.

(v) $(X, \mathbb{F})$ *is a right* MARKOV *process in* $(R, \mathcal{B}^u(R)))$.

(vi) $(X, \mathbb{F}_+)$ *possesses the strong* MARKOV *property.*

COROLLARY 1 *If* $(X, \mathbb{G}_+)$ *possesses the strong* MARKOV *property then* $(X, \mathbb{F})$ *admits a* BOREL *semigroup* $(P_t)_{t\geq 0}$ *of transition probabilities, i.e.,* $(P_t)_{t\geq 0}$ *leaves the space* $\mathbf{B}$ *of bounded* BOREL *functions invariant. In particular, this is true if* $(X, \mathbb{F})$ *is a right* MARKOV *process in* $(R, \mathcal{B}^u(R))$.

The statement of Corollary 1 is the crucial point in the proof of Theorem 2. Actually, we will show a little bit more in the next proposition. Again we do not assume that $(X, \mathbb{F})$ admits a standard semigroup $(P_t)_{t\geq 0}$.

PROPOSITION 2 *Suppose that the following conditions are satisfied:*

(i) *For every $x \in R$, $\mathbf{P}_x(U(x+) > 0)$ and $\mathbf{P}_x(U(x-) > 0)$ only take the values* 0 *or* 1.

(ii) *If $x \in K_+$ (resp. $x \in K_-$) then*

$$\mathbf{P}_x(U(x-) = +\infty) = 1 \quad (\textit{resp.} \quad \mathbf{P}_x(U(x+) = +\infty) = 1).$$

(iii) *$(X, \mathbb{G})$ has the strong* MARKOV *property.*

Let $\mathbf{C}_0$ (resp. $\mathbf{C}_0^u$) be the space of bounded $\mathcal{B}(R)$-measurable (resp. $\mathcal{B}^u(R)$-measurable) finely continuous functions. Then $(P_t)_{t\geq 0}$ leaves $\mathbf{C}_0$ and $\mathbf{C}_0^u$ invariant.

The proof of this proposition will be divided into several steps. Suppose that the conditions of Proposition 2 are satisfied.

LEMMA 1 *For every bounded universally measurable finely continuous function f, the function $P_t f$ is finely continuous.*

Proof In view of assumption (i) in Proposition 2, the states $x \in R$ can be classified as in Definition 2. The structure of the fine topology given in Theorem 1 remains valid. We define

$$U(a) = \inf\{t \geq 0 : X_t = a\}, \quad a \in R. \tag{2}$$

It is easy to see that $U(a)$ is a $\mathbb{G}$-stopping time for every $a \in R$.

Let now f be a bounded universally measurable finely continuous function and (x_n) a decreasing sequence of real numbers converging to $x \in R$. Suppose $x \in K_-^c$. In view of the strong MARKOV property of $(X, \mathbb{G})$ and the continuity of X we get¶

$$\mathbf{E}_x\, I_{\{U(x_n)<+\infty\}}\, \mathbf{E}_{x_n} f(X_t) \;=\; \mathbf{E}_x\, I_{\{U(x_n)<+\infty\}}\, f(X_{U(x_n)+t}). \tag{3}$$

Because of $x \in K_-^c$

$$U(x_n) \downarrow U(x+) = 0 \qquad \mathbf{P}_x\text{-a.s.}$$

¶ I_A denotes the indicator function of the set A.

Since f is finely continuous the (simple) MARKOV property implies

$$\lim_{s \downarrow 0} f(X_{t+s}) = f(X_t) \qquad \mathbf{P}_x\text{-a.s.}$$

(cf. [3], Proposition IV.28) and hence

$$\lim_n f(X_{U(x_n)+t}) = f(X_t) \qquad \mathbf{P}_x\text{-a.s.}$$

Using LEBESGUE's dominated convergence theorem, from (3) we conclude

$$\lim_n \mathbf{E}_{x_n} f(X_t) = \mathbf{E}_x f(X_t)$$

and hence

$$\lim_n P_t(x_n, f) = f(x). \tag{4}$$

Analogously we can show that for any increasing sequence (x_n) converging to $x \in K_+^c$ the relation (4) holds, too. Using Corollary 2 of Theorem 1 we establish that $P_t f$ is, indeed, finely continuous. □

LEMMA 2 *$x \in K_+$ (resp. $x \in K_-$) if and only if for every bounded continuous function f vanishing on $[x, +\infty)$ (resp. on $(-\infty, x]$) and for every $t \geq 0$*

$$\mathbf{E}_x f(X_t) = 0. \tag{5}$$

Proof Because of condition (ii) of Proposition 2 the necessity of (5) is obvious. Conversely, if (5) is satisfied for all bounded continuous functions f vanishing on $[x, +\infty)$ then

$$X_t \geq x \qquad \mathbf{P}_x\text{-a.s.}$$

for all $t \geq 0$. The continuity of the paths of X implies $\mathbf{P}_x(U(x-) = +\infty) = 1$ and hence $x \in K_+$. □

LEMMA 3 (i) *The set K_- (resp. K_+) is closed in the right (resp. left) topology.*

(ii) *K_- and K_+ are BOREL subsets of R.*

(iii) *$K_s = K_- \cup K_+$ is closed.*

Proof Let (x_n) be an increasing sequence from K_+ converging to $x \in R$. Suppose $x \in K_+^c$. Let f be a bounded continuous function vanishing on $[x, +\infty)$. In view of Lemma 1 and Corollary 2 to Theorem 1 we have

$$\mathbf{E}_x f(X_t) = \lim_n \mathbf{E}_{x_n} f(X_t), \qquad t \geq 0.$$

But $x_n \in K_+$ and, in view of (ii) in Proposition 2, $\mathbf{P}_{x_n}(U(x_n-) = +\infty) = 1$. This implies

$$X_t \geq x_n \qquad \mathbf{P}_{x_n}\text{-a.s.}$$

and hence

$$| \mathbf{E}_{x_n} f(X_t) | \leq \sup_{y \in [x_n, x]} | f(y) | .$$

Since $f(x) = 0$ the right member converges to 0. Consequently,

$$\mathbf{E}_x f(X_t) = 0.$$

By Lemma 2, $x \in K_+$. This contradicts the assumption $x \in K_+^c$. Thus $x \in K_+$ and hence K_+ is closed in the left topology. Analogously, it can be shown that K_- is closed in the right topology. This proves (i).

Statement (ii) follows from (i): Let us consider, for example, K_+. Let (x_n) be a sequence which is dense in K_+ and contains all points from K_+ being isolated in the left topology. Then

$$K_+ = \bigcap_{k=1}^{\infty} \bigcup_{j=1}^{\infty} [x_j, x_j + k^{-1}).$$

Finally, we prove (iii). Let (x_n) be a sequence from $K_- \cup K_+$ converging to $x \in R$. First suppose that (x_n) is increasing. If there is a subsequence of (x_n) in K_+ then from (i) we conclude that $x \in K_+ \subseteq K_- \cup K_+$. Hence we can assume that $x_n \in K_-$ for all n sufficiently large. We now show that $x \in K_- \cup K_+$. For this we assume that $x \in K_+^c$. Let f be a bounded continuous function vanishing on $(-\infty, x]$. Using Lemma 1 and Corollary 2 to Theorem 1 we obtain

$$\mathbf{E}_x f(X_t) = \lim_n \mathbf{E}_{x_n} f(X_t).$$

The right hand side is equal to zero because of Lemma 2. Hence we have $\mathbf{E}_x f(X_t) = 0$ for all $t \geq 0$. Again applying Lemma 2 we see that $x \in K_-$. This proves $x \in K_+ \cup K_-$. The case of a decreasing sequence (x_n) is treated analogously. This completes the proof of (iii). □

Proof of Proposition 2 Let f be a bounded $\mathcal{B}(R)$-measurable (resp. $\mathcal{B}^u(R)$-measurable) finely continuous function. According to Lemma 1, $P_t f$ is finely continuous. For completing the proof of the proposition it remains to show that $P_t f$ is $\mathcal{B}(R)$-measurable if f is $\mathcal{B}(R)$-measurable. Let g_- and g_+ be the restrictions of $P_t f$ on K_+^c and K_-^c, respectively. By the fine continuity of $P_t f$, g_- is left continuous and g_+ is right continuous. Hence g_- and g_+ are BOREL functions on K_+^c and K_-^c, respectively. Consequently, there exist BOREL functions G_- and G_+ on R that continue g_- and g_+. Since, because of (ii) in Proposition 2, every $x \in K_- \cap K_+$ is absorbing we get

$$P_t(x, f) = \begin{cases} G_-(x) & \text{if } x \in K_+^c \\ G_+(x) & \text{if } x \in K_-^c \\ f(x) & \text{if } x \in K_- \cap K_+ \end{cases}$$

and using Lemma 3 (ii) we see that $P_t f$ is a BOREL function on R. □

REMARKS 3 (i) If $(X, \mathbb{F})$ is a local martingale with respect to every $\mathbf{P}_x$ then condition (iii) of Proposition 2 that $(X, \mathbb{G})$ has the strong MARKOV property already implies the conditions (i) – (ii) in the formulation of this proposition. See H.J. Engelbert & W. Schmidt [4], Proposition (1.7). Thus in this case the only condition of Proposition 2 is that $(X, \mathbb{G})$ has the strong MARKOV property. This special case was treated in [4], Proposition (1.13).

(ii) For later reference we note that in the proof of Proposition 2 the strong MARKOV property of $(X, \mathbb{G})$ is not exploited in full. The strong MARKOV property is only used for the $\mathbb{G}$-stopping times $U(a)$ of first reaching the level $a \in R$. Consequently, condition (iii) of Proposition 2 can be replaced by the condition that after the time of first reaching the level $a \in R$ the process starts afresh with probability law $\mathbf{P}_a$.

Proof of Theorem 2 The equivalence of (i) and (ii) follows from Proposition 1.

For proving the implication (ii)⇒(iii) we use Theorem IV.26 of [3]. According to this theorem, $(X, \mathbb{F})$ is a right MARKOV process in $(R, \mathcal{B}(R))$ if $(X, \mathbb{F})$ admits a standard semigroup in $(R, \mathcal{B}(R))$, $(X, \mathbb{G})$ is nearly well-measurable in $(R, \mathcal{B}(R))$, and $(X, \mathbb{G})$ satisfies the strong MARKOV property. First we notice that (i) (and (ii)) imply the conditions (i) – (iii) of Proposition 2. Hence $P_t \mathbf{C}_0 \subseteq \mathbf{C}_0$ and, using the monotone class theorem (cf. P.A. Meyer [7], Theorem I.20), we obtain that $(P_t)_{t\geq 0}$ is a BOREL semigroup of transition probabilities, i.e., a semigroup in $(R, \mathcal{B}(R))$. Consequently, if (ii) (and (i)) is satisfied then $(X, \mathbb{F})$ admits a standard semigroup in $(R, \mathcal{B}(R))$. That $(X, \mathbb{F})$ is nearly well-measurable in $(R, \mathcal{B}(R))$ (see [3], No. IV.4) follows from the continuity of the paths of X. Finally, the last condition is contained in (ii).

To prove (iii)$\Rightarrow$(iv) we use Theorem V.33 of [3]. According to this theorem, $(X, \mathbb{F})$ is a perfect MARKOV process with left version $X^- = X$ if $(X, \mathbb{G})$ is nearly previsible in $(R, \mathcal{B}(R))$. This property again follows from the continuity of the paths of X.

The implication (iv)$\Rightarrow$(v) follows from the facts that every perfect MARKOV process in $(R, \mathcal{B}(R))$ is a right MARKOV process in $(R, \mathcal{B}(R))$ and that every right MARKOV process in $(R, \mathcal{B}(R))$ is also a right MARKOV process in $(R, \mathcal{B}^u(R))$ (cf. [3], Corollary IV.24).

The implication (v)$\Rightarrow$(vi) is proven in [3], Theorem IV.6. Finally, (vi)$\Rightarrow$(ii) is trivial. □

One-dimensional continuous Markov processes which are characterized by Theorem 2 have a lot of nice properties. We collect a few of these in the next proposition.

PROPOSITION 3 *Let $(X, \mathbb{F})$ be a continuous Markov process satisfying one (and therefore all) of the conditions (i) – (vi) in Theorem 2. We then have:*

(i) $(X, \mathbb{F})$ *possesses a* FELLER *semigroup of transition probabilities in the fine topology $\mathcal{C}_0$.*

(i) $(X, \mathbb{F})$ *has a* BOREL *semigroup of transition probabilities.*

(iii) $\mathbb{G} = \mathbb{G}_+$.

(iv) *Every $\mathbb{G}$-stopping time T is $\mathbb{G}$-previsible and*

$$\mathcal{G}_{T-} = \mathcal{G}_T = \mathcal{G}_{T+}.$$

Proof Statement (i) means that $P_t \mathbf{C}_0 \subseteq \mathbf{C}_0$ for all $t \geq 0$ and is proven in Proposition 2, (ii) is an immediate consequence of Theorem 2. Property (iii) follows from Proposition 1. Finally, (iv) is a consequence of Theorem 2 (see condition (iv)) and [3], Theorem V.42. □

4 NECESSARY AND SUFFICIENT CONDITIONS

Let $(X, \mathbb{F})$ be a continuous MARKOV process with values in $(R, \mathcal{B}^u(R))$.

The aim of the present section is to give necessary and sufficient conditions for $(X, \mathbb{F}_+)$ possessing the strong MARKOV property.

According to Theorem 2 this is equivalent to asking for conditions that $(X, \mathbb{F})$ is a right MARKOV process in $(R, \mathcal{B}^u(R))$.

We recall that we denote the space of bounded $\mathcal{B}(R)$-measurable (resp. $\mathcal{B}^u(R)$-measurable) finely continuous functions by $\mathbf{C}_0$ (resp. $\mathbf{C}_0^u$).

If the events $\{U(x+) > 0\}$ and $\{U(x-) > 0\}$ are $\mathbf{P}_x$-trivial for every $x \in R$, we introduce the space $\hat{\mathbf{C}}$ (resp. $\hat{\mathbf{C}}^u$) of all bounded $\mathcal{B}(R)$-measurable (resp. $\mathcal{B}^u(R)$-measurable) functions that are right continuous at every $x \in K_-^c$ and left continuous at every $x \in K_+^c$.

THEOREM 3 *The following conditions are equivalent:*

(i) $(X, \mathbb{F})$ *is a right* MARKOV *process in* $(R, \mathcal{B}^u(R))$.

(ii) (a) *For all* $t \geq 0$, $P_t \mathbf{C}_0 \subseteq \mathbf{C}_0$ *(resp.* $P_t \mathbf{C}_0^u \subseteq \mathbf{C}_0^u$*).*

(b) *If* $a \in K_+$ *(resp.* $a \in K_-$*) then*

$$X_t \geq X_{t \wedge U(a)} \quad (\textit{resp. } X_t \leq X_{t \wedge U(a)}) \quad \mathbf{P}_x\text{-}a.s.$$

for every $t \geq 0$ *and* $x \in R$.

(iii) (a) *For every* $x \in R$*, the events* $\{U(x+) > 0\}$ *and* $\{U(x-) > 0\}$ *are* $\mathbf{P}_x$*-trivial.*

(b) *For all* $t \geq 0$, $P_t \hat{\mathbf{C}} \subseteq \hat{\mathbf{C}}$ *(resp.* $P_t \hat{\mathbf{C}}^u \subseteq \hat{\mathbf{C}}^u$*).*

(c) *If* $a \in K_+$ *(resp.* $a \in K_-$*) then*

$$X_t \geq X_{t \wedge U(a)} \quad (\textit{resp. } X_t \leq X_{t \wedge U(a)}) \quad \mathbf{P}_x\text{-}a.s.$$

for every $t \geq 0$ *and* $x \in R$.

(iv) (a) *For every* $x \in R$*, the events* $\{U(x+) > 0\}$ *and* $\{U(x-) > 0\}$ *are* $\mathbf{P}_x$*-trivial.*

(b) *If* $x \in K_+$ *(resp.* $x \in K_-$*) then*

$$\mathbf{P}_x(U(x-) = +\infty) = 1 \quad (\textit{resp. } \mathbf{P}_x(U(x+) = +\infty) = 1).$$

(c) $(X, \mathbb{G})$ *has the strong* MARKOV *property.*

If one (and therefore any) of these conditions is satisfied then $(X, \mathbb{F})$ is a right MARKOV *process in the state space $(R, \mathcal{B}(R))$ and, in particular, admits a standard semigroup $(P_t)_{t\geq 0}$ in $(R, \mathcal{B}(R))$.*

Proof Suppose that $(X, \mathbb{F})$ is a right process in $(R, \mathcal{B}^u(R))$. Note that then the conditions (i) – (iii) of Proposition 2 are satisfied. By Proposition 2 we get (i)(a). Condition (ii)(b) easily follows from the strong MARKOV property of $(X, \mathbb{G})$ applied to the $\mathbb{G}$-stopping time $U(a)$ (cf. Theorem 2). This proves (ii).

For proving the implication (ii)⇒(iii), we first notice that $\mathcal{B}(R) = \sigma(\mathbf{C}_0)$ and $\mathbf{C}_0 \subseteq \mathbf{C}_s$. Hence, according to Theorem II.7 of [3], the condition (ii)(a) yields that $(P_t)_{t\geq 0}$ is a standard semigroup in $(R, \mathcal{B}(R))$ (resp. $(R, \mathcal{B}^u(R))$). In view of Corollary 1 to Proposition 1, condition (iii)(a) holds. Because of Theorem 1, $\mathbf{C}_0 = \hat{\mathbf{C}}$ and $\mathbf{C}_0^u = \hat{\mathbf{C}}^u$. This implies (iii)(b), proving (iii).

Next we verify (iii)⇒(iv). The conditions (iv)(a) and (iv)(b) immediately follow from (iii)(a) and (iii)(c). For proving (iv)(c) it is sufficient to establish that $(X, \mathbb{F})$ is a right MARKOV process. To this end we check the necessary and sufficient conditions given by Theorem IV.33 of [3]. First we notice that the condition (iii)(a) enables us to define $\hat{\mathbf{C}}$ and $\hat{\mathbf{C}}^u$ and Theorem 1 shows that $\hat{\mathbf{C}} = \mathbf{C}_0$ and $\hat{\mathbf{C}}^u = \mathbf{C}_0^u$. In view of (iii)(b)

$$P_t\,\mathbf{C}_0 \subseteq \mathbf{C}_0 \quad (\text{resp.} \quad P_t\,\mathbf{C}_0^u \subseteq \mathbf{C}_0^u) \tag{6}$$

for all $t \geq 0$. Secondly, for the proof of Lemma 3 we only needed (iv)(a), (iv)(b), and (6). Hence $K := K_s = K_+ \cup K_-$ is closed. Let D_K be the first entry time of X into K. Then

$$X_t \geq X_{t\wedge D_K} \qquad on\ \{X_{D_K} \in K_+\} \qquad \mathbf{P}_x\text{-a.s.} \tag{7}$$

and

$$X_t \leq X_{t\wedge D_K} \qquad on\ \{X_{D_K} \in K_-\} \qquad \mathbf{P}_x\text{-a.s.} \tag{8}$$

for every $x \in R$. Indeed, this is obvious from (iii)(c) if $x \in K$. Now suppose $x \in K^c$. Let (a, b) be the component of the open set K^c containing x. Then $D_K = U(a) \wedge U(b)$ $\mathbf{P}_x$-a.s. and the relations (7) and (8) now follow from (iii)(c). From the continuity of the paths of X, the characterization of the fine topology $\mathcal{C}_0$ (see Theorem 1), and the relations (7) and (8) it can easily be derived:

For every $f \in \mathbf{C}_0^u$, $f \circ X$ is right continuous $\mathbf{P}_x$-a.s. for all $x \in R$ (9)

Thirdly, we recall that

$$\mathcal{B}(R) = \sigma(\mathbf{C}_0) \qquad (\text{resp.} \qquad \mathcal{B}(R) \subseteq \sigma(\mathbf{C}_0^u)) \tag{10}$$

since $\mathbf{C} \subseteq \mathbf{C}_0 \subseteq \mathbf{C}_0^u$ by Theorem 1. According to Theorem IV.33 of [3], the conditions (6), (9), and (10) ensure that $(X, \mathbb{F})$ is a right MARKOV process in $(R, \mathcal{B}(R))$(resp. $(R, \mathcal{B}^u(R))$).

Finally, let (iv) be satisfied. Using Proposition 2 we get that $P_t\,\mathbf{C}_0 \subseteq \mathbf{C}_0$ for all $t \geq 0$. As in the proof of the implication (ii)$\Rightarrow$(iii) we observe that the semigroup $(P_t)_{t\geq 0}$ is standard. Together with the condition (iv) this implies that $(X, \mathbb{F})$ is a right MARKOV process in $(R, \mathcal{B}(R))$ as well as in $(R, \mathcal{B}^u(R))$, by Theorem 2. Hence (i) is verified.

The last part of the theorem also follows from Theorem 2.

This completes the proof of the theorem. □

COROLLARY 1 *Let $(X, \mathbb{F})$ be a continuous* MARKOV *process in the state space $(R, \mathcal{B}^u(R))$. Then $(X, \mathbb{F})$ is a right* MARKOV *process in $(R, \mathcal{B}(R))$ if and only if the following conditions are satisfied:*

(a) *For every $x \in R$, $\mathbf{P}_x(U(x+) > 0)$ and $\mathbf{P}_x(U(x-) > 0)$ only take the values 0 or 1.*

(b) *If $x \in K_+$ (resp. $x \in K_-$) then*

$$\mathbf{P}_x(U(x-) = +\infty) = 1 \quad (\textit{resp.} \quad \mathbf{P}_x(U(x+) = +\infty) = 1).$$

(c) *$(X, \mathbb{G})$ has the strong* MARKOV *property at the times $U(a)$ of first reaching the level $a \in R$.*

Proof Since $U(a)$ is a $\mathbb{G}$-stopping time for every $a \in R$ the necessity of the conditions (a) – (c) follows from Theorem 3 (cf. (iv)).

The converse is proven exactly as in the proof of Theorem 3, (iv)$\Rightarrow$(i). For this it suffices to note that in the proof of Proposition 2 the strong MARKOV property of $(X, \mathbb{G})$ was only used for the stopping times $U(a)$, $a \in R$ (see Remarks 3 (ii)). □

COROLLARY 2 *Let $(X, \mathbb{F})$ be a continuous* MARKOV *process in the state space $(R, \mathcal{B}^u(R))$. Assume that $(X, \mathbb{F})$ is regular, i.e.,*

$$\mathbf{P}_x(U(x+) = U(x-) = 0) = 1 \quad \textit{for every} \quad x \in R.$$

Then the following conditions are equivalent:

(i) $(X, \mathbb{F})$ *is a right* MARKOV *process in* $(R, \mathcal{B}^u(R))$.

(ii) *For all* $t \geq 0$, $P_t \mathbf{C} \subseteq \mathbf{C}$ *where* $\mathbf{C}$ *is the space of bounded continuous functions.*

(iii) $(X, \mathbb{G})$ *has the strong* MARKOV *property.*

(iv) *After the time* $U(a)$ *of first reaching the level* $a \in R$ *the process starts afresh with probability law* $\mathbf{P}_a$.

If one (and therefore all) of these conditions is satisfied then $(X, \mathbb{F})$ *is a right* MARKOV *process in* $(R, \mathcal{B}(R))$ *and, in particular, admits a standard semigroup* $(P_t)_{t \geq 0}$ *of transition probabilities in* $(R, \mathcal{B}(R))$.

Proof If $(X, \mathbb{F})$ is regular then, obviously, $\mathbf{C}_0 = \mathbf{C}$. Thus Corollary 2 immediately follows from Theorem 3 and Corollary 1. □

In particular, Corollary 2 states that a regular continuous MARKOV process in $(R, \mathcal{B}^u(R))$ is a right MARKOV process if and only if it is a FELLER process (in the usual topology of the real line).

The following example shows that, in general, condition (ii)(a) of Theorem 3 does not imply the condition (ii)(b).

EXAMPLE 1 Let $(W, \mathbb{F})$ be a BROWNian motion, defined on the family $(\Omega, \mathcal{F}, \mathbf{P}_x, x \in R)$ of probability spaces, with $\mathbf{P}_x(W_0 = x) = 1$, $x \in R$. Let $(Q_t)_{t \geq 0}$ denote the semigroup of $(W, \mathbb{F})$. Now we set

$$X_t = \begin{cases} 0 & \text{if } W_0 = 0 \\ W_t & \text{otherwise} \end{cases}$$

for all $t \geq 0$. Then $(X, \mathbb{F})$ is a continuous MARKOV process in $(R, \mathcal{B}(R))$ with the semigroup $(P_t)_{t \geq 0}$ defined by

$$P_t(x, f) = \begin{cases} f(0) & \text{if } x = 0 \\ Q_t(x, f) & \text{otherwise} \end{cases}$$

for every bounded $\mathcal{B}(R)$-measurable f and $t \geq 0$. Obviously, $K_- = K_+ = \{0\}$. A bounded $\mathcal{B}(R)$-measurable function f belongs to $\mathbf{C}_0$ if and only if f is continuous on $R \setminus \{0\}$. It is now easy to see that $(P_t)_{t \geq 0}$ leaves $\mathbf{C}_0$ invariant. However, condition (ii)(b) of Theorem 3 is not satisfied. Thus $(X, \mathbb{F})$ is not a right MARKOV process. If we set

$$\tilde{X}_t = \begin{cases} 0 & \text{if } W_0 = 0 \\ W_{t \wedge U(0)} & \text{otherwise} \end{cases}$$

then $(\tilde{X}, \mathbb{F})$ becomes a right MARKOV process, with another semigroup than $(P_t)_{t\geq 0}$.

REMARK 4 Let $(X, \mathbb{F})$ be a continuous MARKOV process in the state space $(R, \mathcal{B}^u(R))$ such that $(X, \mathbb{F})$ is a local martingale with respect to every $\mathbf{P}_x$. As proven in [4] (Proposition (1.7)) then the condition (iv)(a) in the formulation of Theorem 3 already follows from (iv)(c). Hence, in this case, $(X, \mathbb{F})$ is a right MARKOV process in $(R, \mathcal{B}(R))$ if and only if $(X, \mathbb{G})$ has the strong MARKOV property (see [4], Theorem (1.17)).

Furthermore, because of $\mathbf{P}_x(U(x+) = U(x-)) = 1$ for every $x \in R$ (cf. [4], proof of Proposition (1.7)), the conditions (a) and (b) of Corollary 1 to Theorem 3 are equivalent to the condition that every holding point is absorbing. (In general, a point $x \in R$ is called a *holding point* if

$$\mathbf{P}_x(U(x+) \wedge U(x-) > 0) > 0.$$

It is said to be *absorbing* if

$$\mathbf{P}_x(U(x+) = U(x-) = +\infty) = 1.)$$

Thus $(X, \mathbb{F})$ is a right MARKOV process if and only if every holding point is absorbing and $(X, \mathbb{G})$ satisfies the strong MARKOV property at every time $U(a)$ of first reaching the level a for all $a \in R$. This result was stated in [4], Theorem (1.20).

5 EQUIVALENT PROCESSES

Let $(X, \mathbb{F})$ be a continuous MARKOV process in the state space $(R, \mathcal{B}^u(R))$ defined on $(\Omega, \mathcal{F}, \mathbf{P}_x, x \in R)$. Let $(P_t)_{t\geq 0}$ be the semigroup of transition probabilities of $(X, \mathbb{F})$.

Suppose that there is given another MARKOV process $(\tilde{X}, \tilde{\mathbb{F}})$ in the state space $(R, \mathcal{B}^u(R))$ defined on the family $(\tilde{\Omega}, \tilde{\mathcal{F}}, \tilde{\mathbf{P}}_x, x \in R)$, admitting the semigroup $(\tilde{P}_t)_{t\geq 0}$ of transition probabilities. We shall assume that $\tilde{X}$ is a separable stochastic process with respect to every $\tilde{\mathbf{P}}_x$. This is, for example, satisfied if $\tilde{X}$ has right (or left) continuous paths.

The processes $(X, \mathbb{F})$ and $(\tilde{X}, \tilde{\mathbb{F}})$ are called equivalent if their semigroups of transition probabilities coincide, i.e., if $P_t = \tilde{P}_t$ for all $t \geq 0$. Obviously, equivalent processes have the same finite dimensional distributions.

THEOREM 4 *Suppose that* $(X, \mathbb{F})$ *is a right* MARKOV *process and that* $(X, \mathbb{F})$ *and* $(\tilde{X}, \tilde{\mathbb{F}})$ *are equivalent. Then* $(\tilde{X}, \tilde{\mathbb{F}})$ *is also a right* MARKOV *process.*

COROLLARY 1 *Assume that* $(X, \mathbb{F})$ *and* $(\tilde{X}, \tilde{\mathbb{F}})$ *are equivalent. If* $(X, \mathbb{G}_+)$ *possesses the strong* MARKOV *property then* $(\tilde{X}, \tilde{\mathbb{F}}_+)$ *also possesses the strong* MARKOV *property.*

COROLLARY 2 *Let* $(P_t)_{t \geq 0}$ *be a semigroup of transition probabilities in* $(R, \mathcal{B}^u(R))$. *Assume that there exists a right* MARKOV *process* $(X, \mathbb{F})$ *admitting* $(P_t)_{t \geq 0}$ *as its semigroup. Then every separable* MARKOV *process* $(\tilde{X}, \tilde{\mathbb{F}})$ *with the semigroup* $(P_t)_{t \geq 0}$ *is a right* MARKOV *process.*

Corollary 1 and Corollary 2 immediately follow from Theorem 4 and Theorem 2.

Proof of Theorem 4 First we notice that the process $\tilde{X}$ has continuous paths $\mathbf{P}_x$-a.s. for every $x \in R$ since $\tilde{X}$ is separable and the equivalent process X is continuous (cf. J. Neveu [8], Proposition III.5.1 and Proposition III.4.1). Therefore we can, and do, assume that $\tilde{X}$ is continuous everywhere. To prove that $(\tilde{X}, \tilde{\mathbb{F}})$ is a right MARKOV process we verify the condition (ii) of Theorem 3. For this we notice that the semigroup $(P_t)_{t \geq 0}$ of $(\tilde{X}, \tilde{\mathbb{F}})$ is a standard semigroup as the semigroup of the right MARKOV process $(X, \mathbb{F})$ (see Theorem 2). The set of right singular points K_+ and of left singular points K_- only depend on the semigroup $(P_t)_{t \geq 0}$ and hence are the same for $(X, \mathbb{F})$ and $(\tilde{X}, \tilde{\mathbb{F}})$. Using Theorem 1 we observe that the fine topology $\mathcal{C}_0$ and, consequently, the space $\mathbf{C}_0$ of bounded $\mathcal{B}(R)$-measurable finely continuous functions is the same for the processes $(X, \mathbb{F})$ and $(\tilde{X}, \tilde{\mathbb{F}})$. Since $(X, \mathbb{F})$ is a right MARKOV process, by Theorem 3 we obtain $P_t \, \mathbf{C}_0 \subseteq \mathbf{C}_0$ for every $t \geq 0$ and hence the condition (ii)(a) for $(\tilde{X}, \tilde{\mathbb{F}})$ of Theorem 3 is satisfied. Also it can easily be seen that the condition (ii)(b) of Theorem 3 is a property of the finite dimensional distributions. It is satisfied for $(X, \mathbb{F})$ and hence for $(\tilde{X}, \tilde{\mathbb{F}})$, too. □

Note that Theorem 4 and its corollaries are no longer true without the condition that the process $\tilde{X}$ is separable. We give the following simple example.

EXAMPLE 2 Let $(X, \mathbb{F})$ be a one-dimensional BROWNian motion, defined on $(\Omega, \mathcal{F}, \mathbf{P}_x, x \in R)$, starting at x with respect to $\mathbf{P}_x$. Let

$a \in R, \;\; a \neq 0$, be fixed. We define

$$\tilde{X}_t = \begin{cases} X_t & \text{if } t \neq U(a) \\ 0 & \text{otherwise.} \end{cases}$$

Since $U(a)$ is continuously distributed, $(\tilde{X}, \mathbb{F})$ is a MARKOV process with the same semigroup as $(X, \mathbb{F})$. However, $(\tilde{X}, \mathbb{F})$ is not a right MARKOV process since it does not satisfy the strong MARKOV property at the stopping time $U(a)$.

Summarizing the results of the present section we can say that the property of a continuous MARKOV process in $(R, \mathcal{B}^u(R))$ to be a right MARKOV process is completely determined by its semigroup of transition probabilities.

REFERENCES

1. R.M. Blumenthal and R.K. Getoor: *Markov Processes and Potential Theory.* New York – London 1968

2. E.B. Dynkin: *Markov Processes.* Fizmatgiz, Moscow 1963

3. H.J. Engelbert: Markov Processes in General State Spaces. Part I, *Math. Nachr.* **80** (1977), 19–36; Part II, *Math. Nachr.* **82** (1978), 191–203; Part III, *Math. Nachr.* **83** (1978), 47–71; Part IV, *Math. Nachr.* **84** (1978), 277–300; Part V, *Math. Nachr.* **85** (1978), 111–130; Part VI, *Math. Nachr.* **85** (1978), 235–266; Part VII, *Math. Nachr.* **86** (1978), 67–83

4. H.J. Engelbert and W. Schmidt: Strong Markov Continuous Local Martingales and Solutions of One-Dimensional Stochastic Differential Equations. Part I, *Math. Nachr.* **143** (1989), 67–184; Part II, *Math. Nachr.* **144** (1989), 241–281; Part III, *Math. Nachr.* **151** (1991), 149–197

5. R.K. Getoor: *Markov Processes: Ray Processes and Right Processes.* Lecture Notes in Mathematics **440**, Springer-Verlag, Berlin 1975

6. K. Itô and H.P. McKean: *Diffusion Processes and Their Sample Paths.* Springer-Verlag, Berlin 1965

7. P.A. Meyer: *Probabilités et potentiel.* Hermann, Paris, 1966

8. J. Neveu: *Bases mathématiques du calcul des probabilités.* Masson et Cie, Paris 1964

9. M.J. Sharpe: *The General Theory of Markov Processes.* Academic-Press, New York 1988

COMPARISON RESULTS FOR STOCHASTIC DIFFERENTIAL EQUATIONS

CHRISTEL GEISS and RALF MANTHEY
Friedrich-Schiller-Universität, Mathematische Fakultät
UHH, 17.OG., D-O-6900 Jena, Germany

Abstract We introduce new comparison theorems for systems of ordinary stochastic differential equations and inequalities as well as for stochastic partial differential equations.

0. INTRODUCTION

Comparison results are a powerful tool both in the theory of ordinary and partial differential equations.[14] For this reason, one looks for analogous statements for stochastic differential equations. Indeed, because of the nondifferentiability of the solution process it is much more difficult to prove such assertions.

A first comparison theorem for ordinary stochastic differential equations was published by Skorohod[12] in 1961. As a consequence of the enormous progress in stochastic analysis his result was generalized and extended by T. Yamada[13], P. Malliavin[7], T. Shiga[5], and many others. Usually, the nondifferentiability of solutions to stochastic differential equations can be compensated in some sense by Itô's formula. Therefore, it is not surprising that most of the above mentioned authors used this tool to prove their comparison results. But one can also establish such results without using Itô's famous formula. We demonstrated this in our joint paper.[11] Later we learned from Bandle and Dozzi[2] about Anderson's paper[1] from 1972. In some sense he used a method which is similar to Skorohod's approach.[12] Both Skorohod and Anderson do not use Itô's formula. Instead of this machinery they work with martingale properties of the stochastic integrals in their setting.

In our case the crucial point is the fact that under certain conditions a continuous martingale can be pathwise differentiated at t=0.

A first comparison result for stochastic partial differential equations was formulated by the second author[8]. Analogous statements were proved by Buckdahn and Pardoux[3], Maslowski[10] and the second author[9]. In these papers equations with an additive (i.e., state independent) noise were considered. In view of comparison results the multiplicative case was first treated by Kotelenez[6] (1989) and by the authors[11]. Analogous results were obtained by Donati-Martin and Pardoux[4] (1992) . A new step forward was done by Bandle and Dozzi[2] (1992). They considered first stochastic differential inequalities. Moreover, their conditions on the coefficients are much more general as those of the authors mentioned above.

We shall consider here systems of (ordinary) stochastic differential equations and inequalities as well as stochastic partial differential equations. The fact that we investigate systems and not only a single equation is a progress in comparison to Bandle and Dozzi. On the other hand, we have different conditions on the coefficients. In particular, we treat situations not considered in their paper.

1. DIFFERENTIABILITY OF MARTINGALES AND COMPARISON RESULTS FOR SYSTEMS OF STOCHASTIC DIFFERENTIAL INEQUALITIES

Let $(\Omega, \mathcal{F}, \mathbb{P})$ be a complete probability space and $\mathbb{F} = (\mathcal{F}_t)_{t\geq 0}$ a right continuous filtration in $\mathcal{F}$ containing all $\mathbb{P}$-null sets of $\mathcal{F}$. In our case the following assertion is a substitute for Itô's formula which can not be used here. But it seems to be of a separate interest, too.

LEMMA 1 *Let $(M, \mathbb{F})$ be a continuous martingale on $(\Omega, \mathcal{F}, \mathbb{P})$ and assume that there exist positive constants c and ε such that*

$$\mathbb{E}\langle M\rangle_t \leq c\, t^2 \cdot |\log t|^{-(1+\varepsilon)}, \qquad t \in [0,1).$$

Then it holds

$$\lim_{t\downarrow 0} t^{-1} \cdot M_t = 0 \qquad \mathbb{P}\text{-}a.s.$$

Consider now the inequalities

$$X_j(t) \le X_j(0) + \int_0^t a_j(s, X(s))ds + \sum_{k=1}^{r} \int_0^t \sigma_{jk}(s, X(s))dW_k(s), \qquad (1A)$$

and

$$Y_j(t) \ge Y_j(0) + \int_0^t b_j(s, Y(s))ds + \sum_{k=1}^{r} \int_0^t \sigma_{jk}(s, Y(s))dW_k(s), \qquad (1B)$$

$$j = 1, ..., d, \quad t \ge 0,$$

where $W = (W_1(t), ...W_r(t))_{t\ge 0}$ is an r-dimensional $\mathbb{F}$-adapted standard Wiener process on $(\Omega, \mathcal{F}, \mathbb{P})$. We shall always assume that the coefficients $a = (a_1, ...a_d)$, $b = (b_1, ..., b_d)$ and $\sigma = (\sigma_{jk})_{\substack{j=1,...,d \\ k=1,...,r}}$ are continuous mappings defined on $\mathbb{R}_+ \times \mathbb{R}^d$.

THEOREM 1 *Suppose that the following two conditions hold.*

$(C1)$ *For any $s \in [0, \infty)$ and $j = 1, ..., d$ the inequality*

$$a_j(s, x) < b_j(s, y)$$

is fulfilled, whenever $x_j = y_j$ and $x_l \le y_l$ for all $l \ne j$.

$(C2)$ *For $\beta \in (\frac{1}{2}, \infty)$ and every $N > 0$ there exist constants K_N such that for $j = 1, ..., d$ it holds*

$$\sum_{k=1}^{r} |\sigma_{jk}(s, x) - \sigma_{jk}(s, y)| \le K_N \cdot |x_j - y_j|^\beta$$

for all $s \in [0, N]$ and $x, y \in \mathbb{R}^d$ with $\|x\|, \|y\| \le N$.

Then, for any local solutions X and Y to the inequalities $(1A)$ and $(1B)$, respectively, on $(\Omega, \mathcal{F}, \mathbb{P})$ with respect to $\mathbb{F}$ and $(W, \mathbb{F})$ it holds $\mathbb{P}$-a.s.

$$X_j(t) \le Y_j(t)$$

for all $t \in [0, \vartheta^X \wedge \vartheta^Y)$, $j = 1, ..., d$.

(We use $\|z\| = \sum_{i=1}^{d} |z_i|$, $z \in \mathbb{R}^d$, and denote the explosion time of a process Z by ϑ^Z.)

Let us shortly sketch the main idea of the proof. First we can reduce the consideration to bounded processes. From $(1A)$ and $(1B)$ we obtain

$$\begin{aligned} 0 &\leq X_j(\tau_j + s) - Y_j(\tau_j + s) \\ &\leq \int_{\tau_j}^{\tau_j+s} [a_j(u, X(u)) - b_j(u, Y(u))]du \qquad (2) \\ &\quad + \sum_{k=1}^{r} \int_{\tau_j}^{\tau_j+s} [\sigma_{jk}(u, X(u)) - \sigma_{jk}(u, Y(u))dW_k(u), \end{aligned}$$

for small $s \in [0, \sigma_j)$, $\sigma_j(\omega) > 0$ on $\{\tau_j < T\}$, $\mathbb{P}$-a.s., where

$$\tau_j := inf\{t > 0 : \ X_j(t) > Y_j(t)\} \wedge T.$$

Using Lemma 1 and some Gronwall type lemmas one obtains from (2)

$$\begin{aligned} 0 &\geq \liminf_{t \downarrow 0} [Y_j(t + \tau_j) - X_j(t + \tau_j)] \\ &\geq b_j(\tau_j, Y_j(\tau_j)) - a_j(\tau_j, X_j(\tau_j)) > 0, \end{aligned}$$

$\mathbb{P} - a.s.$ on $\{\tau_j < T\}, j = 1, ..., d$. This contradiction proves the assertion.

2. COMPARISON RESULTS FOR SYSTEMS OF ORDINARY STOCHASTIC DIFFERENTIAL EQUATIONS

Using Itô's formula one can show the following comparison result for systems of ordinary stochastic differential equations which is quite analogous to the one-dimensional case. Let us consider equations in $(1A)$ and $(1B)$ instead of inequalities.

THEOREM 2 *Suppose that* $(C1)$ *and the following condition are fulfilled.*

(C2) There exists a strictly increasing function* $\rho : \mathbb{R}_+ \to \mathbb{R}_+$ *with* $\rho(0) = 0$ *and*

$$\int_{0+} \rho^{-2}(u)du = \infty$$

such that for each $j = 1, ..., d$

$$\sum_{k=1}^{r} |\sigma_{jk}(s, x) - \sigma_{jk}(s, y)| \leq \rho(|x_j - y_j|)$$

for all $s \in [0, \infty), \quad x, y \in \mathbb{R}^d$.

Suppose also that X, Y solve the systems $(1A), (1B)$, respectively, as equations rather than inequalities. Then

$$X_j(0) \leq Y_j(0) \quad \mathbb{P}\text{-}a.s. \;\; j = 1, ..., d$$

implies

$$X_j(t) \leq Y_j(t) \qquad \mathbb{P}\text{-}a.s.$$

for all $t \in [0, \vartheta^X \wedge \vartheta^Y)$, $j = 1, ..., d$.

The proof uses componentwise the method well-known for the one-dimensional case.[5]

Now we shall give a condition which makes it possible to replace '$<$' in $(C1)$ by '$\leq$'. For this we introduce the following

DEFINITION 1 *A mapping $f : \mathbb{R}^d \to \mathbb{R}^d$ is called quasi-monotonously increasing if for $j = 1, ..., d$*

$$f_j(x) \leq f_j(y)$$

whenever $x_j = y_j$ and $x_l \leq y_l$, $l \neq j$.

THEOREM 3 *If the solution X (Y) to the equations $(1A)$ $((1B))$ (i.e., with the inequality sign replaced by equality sign) is pathwise unique and the coefficient a (b) is quasi-monotonously increasing with respect to $x \in \mathbb{R}^d$, then condition $(C1)$ in Theorem 2 can be replaced by the following:*
$(C1^*)$ For $s \in \mathbb{R}_+$, $j = 1, ..., d$, and $x \in \mathbb{R}^d$ it holds

$$a_j(s, x) \leq b_j(s, x).$$

We shall sketch the proof. Let us suppose that X is pathwise unique and a is quasi-monotonously increasing with respect to $x \in \mathbb{R}^d$. It suffices to consider bounded a, b and σ. Choose $\varepsilon > 0$ and put

$$a_j^\varepsilon := a_j - \varepsilon \quad j = 1, \ldots, d,$$

Then we get

$$X_j^\varepsilon(t) \leq Y_j^\varepsilon(t)$$

$\mathbb{P}$-a.s. for all $t \geq 0$, $j = 1, \ldots, d$. Analogously, we observe

$$X_j^{\varepsilon_1}(t) \leq X_j^{\varepsilon_2}(t) \leq \ldots \leq Y_j(t) \;\; \mathbb{P}\text{-a.s.}$$

for a strictly decreasing sequence $(\varepsilon_n)_{n\geq 0}$ with limit zero as well as

$$X_j^{\varepsilon_1}(t) \leq X_j^{\varepsilon_2}(t) \leq \ldots \leq X_j(t) \quad \mathbb{P}\text{-a.s.},$$

$t \geq 0$, $j = 1, \ldots, d$. To complete the proof one can easily verify that $\lim_{n\to\infty} X^{\varepsilon_n}$ (pointwise) is a modification of X.

By analogy with Anderson[1], one proves the following result.

COROLLARY *Suppose that the assumptions of Theorem 2 are satisfied, and let a be quasi-monotonously increasing and locally Lipschitz continuous with respect to $x \in \mathbb{R}^d$. Then the solution X to $(1A)$ is pathwise unique.*

3. A COMPARISON THEOREM FOR A STOCHASTIC PARTIAL DIFFERENTIAL EQUATION

In this section we formulate a comparison theorem for solutions of the following (formal) Dirichlet problem:

$$\begin{aligned} \frac{\partial}{\partial t}u(t,x) &= \frac{1}{2}(\Delta u)(t,x) + f(u(t,x)) + \sigma(u(t,x)) \cdot \frac{\partial^2}{\partial t\, \partial x}W(t,x), \\ & \qquad\qquad\qquad\qquad\qquad\qquad x \in (0,1), \quad t > 0, \\ u(t,0) &= u(t,1) = 0, \qquad\qquad t > 0, \\ u(0,x) &= \varphi(x), \qquad\qquad x \in (0,1). \end{aligned} \tag{D}$$

Here $\frac{\partial^2}{\partial t\,\partial x}W$ stands for a Gaussian space-time white noise.

DEFINITION 2 *By a solution of* (D) *we mean a pathwise continuous random field $u : \mathbb{R}_+ \times [0,1] \times \Omega \to \mathbb{R}$ defined on a complete probability space $(\Omega, \mathcal{F}, \mathbb{P})$ with a right continuous filtration $\mathbb{F} = (\mathcal{F}_t)_{t\geq 0}$ such that*

(i) there exists a Brownian sheet $W : \mathbb{R}_+^2 \times \Omega \to \mathbb{R}$ such that for every $x \in (0,1]$ the process $(W(t,x))_{t\geq 0}$ is a martingale with respect to $\mathbb{F}$,

(ii) $u(t,\cdot)$ is $\mathcal{F}_t$-measurable for any $t \geq 0$,

(iii) u and W satisfy $u(0,x)=\varphi(x)$ *and*

$$\begin{aligned} u(t,x) &= \int_0^1 G(t,x,y)\varphi(y)dy \\ &+ \int_0^t \int_0^1 G(t-s,x,y)f(u(s,y))dy\,ds \qquad \text{(I)} \\ &+ \int_0^t \int_0^1 G(t-s,x,y)\sigma(u(s,y))W(dy\,ds), \end{aligned}$$

$\mathbb{P}$*-a.s. for* $t>0$ *and* $x\in[0,1]$.

Here G denotes the Green's function corresponding to $\frac{\partial}{\partial t}-\frac{1}{2}\triangle$ and Dirichlet boundary conditions. The stochastic integral in (I) can be understood as an integral with respect to a worthy martingale measure in the sense of Walsh[14]. For relations to integrals with respect to a cylindrical (or standard $\mathcal{D}'$) Brownian motion, see Manthey and Stiewe.[11]

Denote the set $\{h\in\mathbb{C}([0,1];\mathbb{R}):\ h(0)=h(1)=0\}$ by $\mathbb{C}_0$.

We need the following two conditions.

$(C3)$ *The mapping* $\sigma:\mathbb{R}\to\mathbb{R}$ *is continuous, bounded and satisfies*

$$|\sigma(u)-\sigma(v)|\le\rho(|u-v|)$$

for all u, $v\in\mathbb{R}$.
(Here ρ is the function defined in $(C2^*)$.)

$(C4)$ *The mappings* $f_1,f_2:\mathbb{R}\to\mathbb{R}$ *are Lipschitz continuous, and the relation*

$$f_1(u)\le f_2(u), \qquad u\in\mathbb{R}$$

is fulfilled.

Let u_1 and u_2 be solutions to ($\mathbb{D}$) corresponding to f_1 and f_2, respectively, instead of f. Suppose that they have initial functions φ_1 and φ_2 both given on $(\Omega,\mathcal{F},\mathbb{F},\mathbb{P})$ with the same Brownian sheet W, and $\varphi_j\in\mathbb{C}_0$ pathwise, $i=1,2$.

THEOREM 4 *Assume* $(C3)$ *and* $(C4)$ *and let* u_i, $i=1,2$, *be unique in the sense of probability law. If the initial functions satisfy*

$$\varphi_1(x)\le\varphi_2(x) \qquad \mathbb{P}\text{-a.s.},\quad x\in[0,1],$$

then it holds

$$u_1(t,x) \leq u_2(t,x) \qquad \mathbb{P}\text{-}a.s.$$

for any $(t,x) \in [0,\infty) \times [0,1]$.

The proof of this assertion is based on Funaki's polygonal approximation.[11] By the use of this approach we obtain a system of ordinary stochastic differential equations to which we can apply the previous results.

ACKNOWLEDGEMENT

The first named author would like to thank the DFG for financial support.

REFERENCES

1. Anderson, W.J., Local behaviour of solutions of stochastic integral equations. *Transactions of the American Mathematical Society* **164** (1972) 309–320.
2. C. Bandle and M. Dozzi, Comparison theorems for stochastic differential inequalities and an application to reaction–diffusion equations with random sources, *preprint* (1992).
3. R. Buckdahn and E. Pardoux, Monotonicity methods for white noise driven SPDE's, in : *Diffusion processes and related problems in Analysis,* Vol. I, M. Pinsky ed., Birkhäuser (1990) 219–233.
4. C. Donati–Martin and E. Pardoux, White noise driven SPDE's with reflection, *preprint* (1992).
5. N. Ikeda and S. Watanabe, *Stochastic differential equations and diffusion processes* North–Holland, Amsterdam (1981).
6. P. Kotelenez, Existence, uniqueness and smoothness for a class of function–valued stochastic partial differential equations, *preprint* (1989).
7. P. Malliavin, *Géométrie differentielle stochastique,* Les Presses de l' Université de Montréal, Montreal (1978).
8. R. Manthey, Existence and uniqueness of a solution of a reaction–diffusion equation with polynomial nonlinearity and white noise disturbance, *Mathematische Nachrichten* **125** (1986) 121–133.
9. R. Manthey, On the Cauchy problem for reaction–diffusion equation with white noise, *Mathematische Nachrichten* **136** (1988) 209–228.
10. R. Manthey and B. Maslowski, Qualitative behaviour of solutions of stochastic reaction–diffusion equations, *Stochastic Processes and their Applications* **42** (1992) 265–289.

11. R. Manthey and C. Stiewe, Existence and uniqueness of solutions to Volterra's population equation with diffusion and noise, *Stochastics and Stochastics Reports* **41** (1992) 135–161.
12. A. V. Skorohod: *Investigations in the theory of random processes,* University of Kiev, Kiev (1961) (in Russian).
13. T. Yamada: On a comparison theorem for solutions of stochastic differential equations and its application, *Journal of Mathematics of Kyoto University,* **11** (1971) 155–167.
14. J. B. Walsh, An introduction to stochastic partial differential equations, *Lecture Notes in Mathematics* **1180** (1986) 266–439.
15. W. Walter, *Differential and Integral Inequalities,* Springer (1970).

ON LUNDBERG INEQUALITIES IN A MARKOVIAN ENVIRONMENT

BRONIUS GRIGELIONIS
Institute of Mathematics and Informatics
Akademijos 4, 2600 Vilnius, Lithuania

Abstract. In this paper we shall develop a martingale approach, initiated by H.U.Gerber in 1, to prove exponential inequalities for the ruin probabilities in an insurance models, when the underlying point process, describing the occurence of the claims, is a Cox process, conditional intensity of which is a function of a Markov process with a general state spase. The solution of this problem is reduced to some spectral problem of the operator, expressed in terms of the infinitesimal operator of the Markov process and the function, defining conditional intensity of the Cox process. Several examples illustrate the general approach.

Let $(\Omega, \mathcal{F}, \mathbb{F}, P)$ be a filtered probability space and $(\mathcal{Y}, \mathcal{B})$ be a measurable space. We shall denote $(\mathbb{B}_{\mathcal{Y}}, ||\cdot||)$ some Banach space of $\mathcal{B}$-measurable functions.

Assume that we are given (i) a sequence $\{Z_k,\ k \geqslant 1\}$ of nonnegative independent identically distributed random variables, describing the costs of the claims, having the common distribution function $F(x)$, such that for some $r_\infty,\ 0 < r_\infty \leqslant \infty$, and each $r < r_\infty$

$$h(r) := \int_0^\infty e^{rx}\, dF(x) - 1 < \infty, \tag{1}$$

(ii) a time homogeneous $(P, \mathbb{F})$ - Markov process $\{Y(t),\ t \geqslant 0\}$ with a state space $(Y, \mathcal{B})$ of environment, an initial distribution $P_0(B)$

and a probability transition function $P(t, y, B)$, $t \geqslant 0$, $y \in \mathcal{Y}$, $B \in \mathcal{B}$, i.e., for each $0 \leqslant s < t$, $B \in \mathcal{B}$ P-a.e.

$$P\{Y(t) \in B \mid \mathcal{F}_s\} = P(t-s, Y(s), B)$$

and

$$P\{Y(0) \in B\} = P_0(B),$$

and (iii) a $\mathbb{F}$-adapted point process $\{N(t),\ t \geqslant 0\}$, describing the occurrence of the claims, independent of $\{Z_k,\ k \geqslant 1\}$, such that for each $w \in R^1$, $0 \leqslant s < t$ P-a.e.

$$E(e^{iw(N(t)-N(s))} \mid \mathcal{F}_s \vee \mathcal{G}) = \exp\Big\{\int_s^t \lambda(Y(v))\, dv(e^{iw} - 1)\Big\}, \qquad (2)$$

where λ is a finite nonnegative $\mathcal{B}$-measurable function; $\mathcal{G} = \sigma\{Y(t),\ t \geqslant 0\}$.

Denote

$$T_t f(y) = \int_{\mathcal{Y}} f(z) P(t, y, dz), \quad t \geqslant 0, \quad y \in \mathcal{Y}, \quad f \in \mathbb{B}_{\mathcal{Y}},$$

assume that T_0 is the identity, for each $t > 0$

$$T_t \mathbb{B}_{\mathcal{Y}} \subseteq \mathbb{B}_{\mathcal{Y}} \qquad (3)$$

and

$$\lim_{s \to 0} ||T_{t+s} f - T_t f|| = 0, \quad f \in \mathbb{B}_{\mathcal{Y}}. \qquad (4)$$

Let $(A, D(A))$ be the infinitesimal operator of the semigroup T_t, $t \geqslant 0$.

Usually we shall consider cases, when $\mathbb{B}_{\mathcal{Y}}$ is a class of all $\mathcal{B}$-measurable functions f, such that $||f|| := \sup_{y \in \mathcal{Y}} |f(y)| < \infty$, or $\mathbb{B}_{\mathcal{Y}} = L^p(\mathcal{Y}, \mathcal{B}, \nu)$, $p \geqslant 1$, where ν is a σ-finite measure on $\mathcal{B}$. Remark that in the last case (3) is always satisfied, when ν is a finite invariant measure, i.e. for each $t > 0$ and $B \in \mathcal{B}$

$$\int_{\mathcal{Y}} P(t, y, B) \nu(dy) = \nu(B) \qquad \text{(see, e.g. 2).}$$

For each $f \in D(A)$ denoting

$$M^f(t) = f(Y(t)) - f(Y(0)) - \int_0^t Af(Y(s))\,ds, \quad t \geqslant 0, \tag{5}$$

assume that M^f is a $(P, \mathbb{F})$-local martingale (for terminology see, e.g., 3).

Taking $Z_0 \equiv 0$, for a positive constant rate of income c define the risk process X by means of the formula:

$$X(t) = ct - \sum_{k=0}^{N(t)} Z_k, \quad t \geqslant 0,$$

and the time of ruin with the initial capital u as the following stopping time:

$$T_u = \inf\{t : u + X(t) < 0\}.$$

We shall consider the functions:

$$\Psi(u) := P\{T_u < \infty\}$$

and

$$\Psi(u,t) := P\{T_u \leqslant t\}, \quad t \geqslant 0, \quad u \geqslant 0.$$

Assume that (i) for some subset $\Gamma_0 \subset [0, \infty)$ and each $\gamma \in \Gamma_0$ there exist $g(\gamma, \cdot\,) \in D(A)$, $g(\gamma, \cdot\,) > 0$ and $\mu(\gamma) \in \mathbb{R}^1$ such that

$$Ag(\gamma, y) + \gamma\lambda(y)g(\gamma, y) = \mu(\gamma)g(\gamma, y), \quad y \in \mathcal{Y}, \tag{6}$$

(ii) for some $R \in (0, r_\infty)$, called the Lundberg exponent, such that $h(R) \in \Gamma_0$,

$$\mu(h(R)) = cR, \tag{7}$$

and (iii) for fixed $\underline{t}$, $\overline{t}$, $0 \leqslant \underline{t} < \overline{t} < \infty$, there exists $r_{\underline{t},\overline{t},} < r_\infty$, such that $h(r_{\underline{t},\overline{t},}) \in \Gamma_0$ and

$$R_{\underline{t}}^{\overline{t}} := \sup_{\substack{0 \leqslant r < r_\infty \\ h(r) \in \Gamma_0}} \min(r - \underline{t}(\mu(h(r)) - cr), r - \overline{t}(\mu(h(r)) - cr)) \tag{8}$$

$$= \min(r_{\underline{t},\overline{t}} - \underline{t}(\mu(h(r_{\underline{t},\overline{t}})) - cr_{\underline{t},\overline{t}}), \quad r_{\underline{t},\overline{t}} - \overline{t}(\mu(h(r_{\underline{t},\overline{t}})) - cr_{\underline{t},\overline{t}})).$$

$R_{\underline{t}}^{\bar{t}}$ is called the finite-time Lundberg exponent.

Proposition 1. (cf. 4, 5). If $Eg_R(Y(0)) < \infty$, under assumptions (1)-(7) for each $u \geqslant 0$ the Lundberg inequality holds:

$$\Psi(u) \leqslant \frac{Eg_R(Y(0))}{E(g_R(Y(T_u)) \mid T_u < \infty)} e^{-Ru}, \tag{9}$$

where $g_r(\cdot) = g(h(r), \cdot)$.

If $Eg_{r_{\underline{t},\bar{t}}}(Y(0)) < \infty$, under assumptions (1)-(6), (8) for each $u \geqslant 0$ the finite-time Lundberg inequality holds:

$$\Psi(u, \bar{t}u) - \Psi(u, \underline{t}u) \leqslant \frac{Eg_{r_{\underline{t},\bar{t}}}(Y_0)}{E(g_{r_{\underline{t},\bar{t}}}(Y(T_u)) \mid \underline{t}u < T_u \leqslant \bar{t}u)} e^{-R_{\underline{t}}^{\bar{t}} u}. \tag{10}$$

Proof. From the above assumption it follows that X is a $(P, \mathbb{F}_{\mathcal{G}})$-semimartingale with the triplet $(\alpha, 0, \Pi)$ of predictable charactersitics, where

$$\alpha(t) = ct + \int_{-1}^{0} x\, dG(x) \cdot \int_0^t \lambda(Y(s))\, ds, \quad t \geqslant 0,$$

$$\Pi([0,t] \times \Gamma) = \int_0^t \Lambda(Y(s))\, ds \int_{\Gamma \cap (-\infty, 0)} dG(x), \quad t \geqslant 0, \quad \Gamma \in \mathcal{B}(R^1/\{0\}),$$

and $G(x) = P\{-Z_1 \leqslant x\}$, $x \in R^1$; $\mathbb{F}_{\mathcal{G}} = \{\mathcal{F}_t \vee \mathcal{G},\ t \geqslant 0\}$.

Thus, according to the criterion in 6, $(X(t),\ Y(t))$, $t \geqslant 0$, is a Markov additive process, i.e. for each $0 \leqslant s < t$, $\Gamma \in \mathcal{B}(R^1)$, $B \in \mathcal{B}$ P-a.e.

$$\begin{aligned} &P\{X(t) - X(s) \in \Gamma,\ \ Y(t) \in B \mid \mathcal{F}_s\} \\ &\quad = P\{X(t) - X(s) \in \Gamma,\ \ Y(t) \in B \mid Y(s)\}. \end{aligned}$$

Easy calculations show that for each $0 \leqslant s < t$ and $r < r_\infty$

$$\begin{aligned} &E[\exp\{-r(X(t) - X(s))\} \mid \mathcal{F}_s \vee \mathcal{G}] \\ &\quad = \exp \int_s^t (\lambda(Y(v))h(r) - cr)\, dv\}. \end{aligned} \tag{11}$$

Define

$$M_u^r(t) := e^{-r(u+X(t))} g_r(Y(t)) e^{-(\mu(h(r))-cr)t}, \quad r < r_\infty, \quad h(r) \in \Gamma_0, \quad t \geqslant 0, \tag{12}$$

and prove the following lemma.

Lemma. If $Eg_r(Y_0) < \infty$, $h(r) \in \Gamma_0$, $r < r_\infty$, then M_u^r is a supermartingale.

Indeed, using Ito's formula, from (6) and (7), we find that

$$\begin{aligned} L^r(t) : &= \exp\Big\{ \int_0^t (\lambda(Y(v))h(r) - \mu(h(r)))\, dv \Big\} g_r(Y(t)) \\ &= g_r(Y(0)) + \int_0^t \exp\Big\{ (\Lambda(Y(v))h(r) \\ &\quad - \mu(h(r)))\, dv \Big\} dM^{g_r}(s), \quad t \geqslant 0. \end{aligned} \tag{13}$$

Thus L^r is a positive $(P, \mathbb{F})$-local martingale. Because $Eg_r(Y(0)) < < \infty$, L^r is a $(P, \mathbb{F})$-supermartingale and therefore, having in mind (11) and (13), for each $0 \leqslant s < t$ we have that P-a.e.

$$\begin{aligned} E(M_u^r(t) \mid \mathcal{F}_s) &= e^{-r(u+X(s))} E(E(\exp\{-r(X(t) \\ &\quad - X(s))\} \mid \mathcal{F}_s \vee \mathcal{G}) g_r(Y(t)) \mid \mathcal{F}_s) e^{-(\mu(h(r))-cr)t} \\ &= e^{-r(u+X(s))} e^{-(\mu(h(r))-cr)s} E\Big(\exp\Big\{ \int_s^t (\lambda(Y(v))h(r) \\ &\quad - \mu(h(r)))\, dv \Big\} g_r(Y(t)) \mid \mathcal{F}_s \Big) \\ &= e^{-r(u+X(s))} e^{-(\mu(h(r))-cr)s} E(L^r(t) \mid \mathcal{F}_s) \exp\Big\{ -\int_0^s (\lambda(Y(v))h(r) \\ &\quad - \mu(h(r))\, dv \Big\} \leqslant e^{-r(u+X(s))} g_r(Y(s)) e^{-(\mu(h(r))-cr)s} = M_u^r(s). \end{aligned}$$

Thus the Lemma is proved.

Applying the optional sampling theorem for supermartingales (see, e.g. 3) and noting that $u + X(T_u) \leqslant 0$ on $\{T_u < \infty\}$, for each $t_0 > 0$ we have the inequalities (cf. 4):

$$\begin{aligned} EM_u^R(0) &\geqslant EM_u^R(T_u \wedge t_0) \\ &\geqslant E[e^{-R(u+X(T_u))} g_R(Y(t_u)) \mid T_u \leqslant t_0] P\{t_u \leqslant t_0\} \\ &\geqslant E[g_R(Y(T_u)) \mid T_u \leqslant t_0] P\{T_u \leqslant t_0\}. \end{aligned} \tag{14}$$

Because

$$M_u^R(0) = g_R(Y(0)) e^{-Ru},$$

taking the limit, as $t_0 \to \infty$, from (14) we derive (9).

Analogously we find that

$$\begin{aligned} & e^{-r_{\underline{t},\bar{t}} u} E g_{r_{\underline{t},\bar{t}}}(Y(0)) = EM_u^{r_{\underline{t},\bar{t}}}(0) \geqslant EM_u^{r_{\underline{t},\bar{t}}}(T_u \wedge \bar{t}u) \\ & \geqslant E(M_u^{r_{\underline{t},\bar{t}}}(T_u) \mid \underline{t}u < T_u \leqslant \bar{t}u) P\{\underline{t}u < T_u \leqslant \bar{t}u\} \\ & \geqslant E(g_{r_{\underline{t},\bar{t}}}(Y(T_u) e^{-(\mu(h(r_{\underline{t},\bar{t}}))-cr_{\underline{t},\bar{t}})T_u} \mid \underline{t}u < T_u \leqslant \bar{t}u) \\ & \times P\{\underline{t}u < T_u \leqslant \bar{t}u\} \\ & \geqslant E(g_{r_{\underline{t},\bar{t}}}(Y(T_u)) \mid \underline{t}u < T_u \leqslant \bar{t}u) \exp\{-\max(\underline{t}(\mu(h(r_{\underline{t},\bar{t}})) \\ & - cr_{\underline{t},\bar{t}}), \bar{t}(\mu(h(r_{\underline{t},\bar{t}})) - cr_{\underline{t},\bar{t}}))u\} P\{\underline{t}u < T_u \leqslant \bar{t}u\} \end{aligned}$$

and thus

$$P\{\underline{t}u < T_u \leqslant \bar{t}u\} \leqslant \frac{E g_{r_{\underline{t},\bar{t}}}(Y(0))}{E(g_{r_{\underline{t},\bar{t}}}(Y(T_u)) \mid \underline{t}u < T_u \leqslant \bar{t}u)} e^{-R_{\underline{t}}^{\bar{t}} u}.$$

Proposition 1 is proved.

Example 1 (classical model). Let $\lambda(y) \equiv \lambda$, $1 \in D(A)$ and $A1 = 0$. Then $\mu(\gamma) = \lambda\gamma$, $\gamma \geqslant 0$, and $g(\gamma, \cdot) = 1$ solve (6). Under conditions that we have for ralative safety loading

$$\rho := \frac{c - \lambda\mu}{\lambda\mu} > 0,$$

and $h(r) \uparrow \infty$ as $r \uparrow r_\infty$, the Lundberg exponent R, satisfying

$$h(R) = \frac{c}{\lambda} R,$$

exists (see, e.g., 1, 4); $\mu = EZ_1$.

Example 2. Let $\mathcal{Y} = [0,1]$, $\mathcal{B} = \mathcal{B}([0,1])$, $P_0(dy) = dy$, $P(t,y,B) = \int\limits_B p(t,y,z)\,dz$, $t \geqslant 0$, $y \in [0,1]$, $B \in \mathcal{B}([0,1])$,

$$p(t,y,z) = \sum_{n=-\infty}^{\infty} \exp\Big\{-\frac{\sigma^2}{2}tn^2\pi^2\Big\} \cos n\pi y \cos n\pi z,$$
$$t \geqslant 0, \quad y,z \in [0,1],$$

$By = L^2([0,1],\mathcal{B}([0,1]),P_0)$, $Af = \frac{\sigma^2}{2}f''$, $D(A) = \{f : f \in B_{\mathcal{Y}}, f$ is differentiable, f' is absolute continuous, $f'' \in \mathbb{B}_{\mathcal{Y}}$, $f'(0) = f'(1) = 0\}$, $\lambda(\cdot)$ be a continuous nonnegative function on $[0, 1]$. Obviously $\{Y(t),\ t \geqslant 0\}$ is the Brownian motion on $[0, 1]$ with the reflecting boundaries and the diffusion coefficient σ^2. The conditional intensity of the occurrence of the claims $\{\lambda(Y(t)),\ t \geqslant 0\}$ is a stationary process with the correlation function

$$r_\lambda(t) = \sum_{n=1}^{\infty} \exp\Big\{-\frac{\sigma^2}{2}n^2\pi^2 t\Big\}a_n^2, \qquad t \geqslant 0,$$

where

$$a_n = \int_0^1 \lambda(y) \cos n\pi y\, dy, \qquad n \geqslant 1.$$

The equation (6) reduces to the classical Sturm Liouville problem:

$$\frac{\sigma^2}{2}g''(\gamma,y) + \gamma\lambda(y)g(\lambda,y) = \mu(\gamma)g(\gamma,y), \qquad y \in (0,1),$$

$$g'(\gamma,0) = g'(\gamma,1) = 0.$$

Choosing $g(\gamma,0) = 1$, it is well known (see, e.g., 7) that $\mu(\gamma)$, $\gamma \geqslant 0$, coincides with the maximal eigenvalue of the operator $A + \gamma\lambda(\cdot)$, $g(\gamma,\cdot)$ is the corresponding strictly positive eigenfunction and the variational principle holds:

$$\mu(\gamma) = \sup_{\substack{f \in D(A) \\ ||f||=1}} [(Af,f) + \gamma(\lambda f,f)].$$

This situation is very typical for the wide class of Markov process Y. For example, under suitable smoothness assumptions for diffusion processes on $\mathcal{Y} = G \cup \partial G \subset R^d$, $d \geqslant 1$, where G is a bounded domain, satisfying the mixed reflection and absorbtion boundary conditions, an extention of the above results is true (see, e.g., 7, ch. V).

Example 3. Let $\mathcal{Y} = \{1, \ldots, M\}$, $A = (a_{jk})_{j,k=1}^{M}$, where a_{jk} denotes the transition intensities, $\lambda(j) \geqslant 0$, $j = 1, \ldots, M$. If the corresponding Markov process $\{Y(t),\ t \geqslant 0\}$ is irreducible, then $\mu(\gamma)$, $\gamma \geqslant 0$, is the eigenvalue of the matrix $(a_{jk} + \gamma\lambda(j)\delta_{jk})_{j,k=1}^{M}$ with the maximal real part and $g(\gamma, \cdot)$ is the corresponding eigenvector with the positive components. This model is well investigated in the case when $\{\lambda(Y(t)),\ t \geqslant 0\}$ is a Markov process (see, e.g., 4, 5, 8-11 and references therein).

Example 4. Let $(\mathcal{Y}, \mathcal{B})$ be an arbitrary measurable space, $\{y\} \in \mathcal{B}$ for each $y \in \mathcal{Y}$, the probability transition function $P(t, y, B)$ satisfies the jump condition:

$$\lim_{t \downarrow 0} P(t, y, \{y\}) = 1, \qquad y \in \mathcal{Y},$$

there exist

$$q(y) := \lim_{t \downarrow 0} (1 - P(t, y, \{y\}))/t < \infty, \qquad y \in \mathcal{Y},$$

$$q(y, B) := \lim_{t \downarrow 0} P(t, y, B \backslash \{y\})/t, \quad y \in \mathcal{Y}, \quad B \in \mathcal{B},$$

and $q(y) = q(y, \mathcal{Y}) > 0$, $y \in \mathcal{Y}$.

Assume that

$$q(y, B) = q(y) Q(B), \quad y \in \mathcal{Y}, \quad B \in \mathcal{B},$$

where Q is a probability measure on $\mathcal{B}$ such that $Q(\{y\}) = 0$ for all y and

$$q_0 := \int_{\mathcal{Y}} \frac{Q(dy)}{q(y)} < \infty.$$

Let

$$P_0(B) = \frac{1}{q_0} \int_{B} \frac{Q(dy)}{q(y)}, \qquad B \in \mathcal{B}.$$

Because for each $B \in \mathcal{B}$

$$\int_{B} q(y) P_0(dy) = \int_{\mathcal{Y}} q(y, B) P_0(dy),$$

the measure P_0 is invariant.

Consider $\mathbb{B}_{\mathcal{Y}} = L^1(\mathcal{Y}, \mathcal{B}, P_0)$, $D(A) = L^1(\mathcal{Y}, \mathcal{B}, P_0) \cap L^1(\mathcal{Y}, \mathcal{B}, Q)$,

$$Af(y) = -q(y)f(y) + q(y) \int_{\mathcal{Y}} f(y)Q(dy), \qquad y \in \mathcal{Y}.$$

For a nonnegative $\mathcal{B}$-measurable function $\lambda(y)$, $y \in \mathcal{Y}$, and a subset $\Gamma_0 \subset [0, \infty)$ assume that for each $\gamma \in \Gamma_0$ there exists $\mu(\gamma) \in R^1$ such that

$$q(y) - \gamma\lambda(y) + \mu(\gamma) > 0 \qquad Q\text{-a.e.},$$

$$\int_{\mathcal{Y}} \frac{q(y)Q(dy)}{q(y) - \gamma\lambda(y) + \mu(\gamma)} = 1$$

and

$$\int_{\mathcal{Y}} \frac{(\lambda(y) + 1)Q(dy)}{q(y) - \gamma\lambda(y) + \mu(\gamma)} < \infty.$$

Under these assumptions the functions

$$g(\gamma, y) = \frac{q(y)}{q(y) - \gamma\lambda(y) + \mu(\gamma)}, \quad y \in \mathcal{Y}, \quad \gamma \in \Gamma_0,$$

solve (6).

When $\mathcal{Y} = [0, \infty)$ and $\lambda(y) = y$, $y \geqslant 0$, this model is analysed in detail in 4, 10.

Example 5. Let $\mathcal{Y} = R^1$, $\mathcal{B} = \mathcal{B}(R^1)$, $P_0(dy) = \sqrt{\frac{\alpha}{\pi\sigma^2}} e^{-\frac{\alpha y^2}{\sigma^2}} dy$, $\alpha > 0$,

$$P(t, y, B) = \int_B p(t, y, z)\, dz, \quad t \geqslant 0, \quad y \in R^1, \quad B \in \mathcal{B}(R^1),$$

$$p(t, y, z) = \sqrt{\frac{\alpha}{\pi(1 - e^{-2\alpha t})\sigma^2}} \cdot$$
$$\cdot \exp\left\{\frac{\alpha}{(1 - e^{-2\alpha t})\sigma^2}(y - ze^{-\alpha t})^2\right\}, \quad t \geqslant 0,$$

$y, z \in R^1$, $\mathbb{B}_{\mathcal{Y}} = L^1(R^1, \mathcal{B}(R^1), P_0)$, $D(A) = \{f : f \in \mathbb{B}_{\mathcal{Y}}$, f is differentiable, f' is absolute continuous,

$$\int_{R^1} |yf'(y)| P_0(dy) < \infty, \quad f'' \in \mathbb{B}_{\mathcal{Y}}\},$$

$$Af(y) = \frac{\sigma^2}{2} f''(y) - \alpha y f'(y), \quad y \in R^1.$$

Thus, $\{Y(t),\ t \geqslant 0\}$ is an Ornstein-Uhlenbeck process. Explicit formulas for $\mu(\gamma)$ and $g(\gamma, \cdot)$ are found in 12, when $\lambda(y) = m + y^2$, $m \geqslant 0$, or $\lambda(y) = (m+y)^2$, $y \in R^1$.

Finally remark that slight changes of arguments permit us to extend the above results to the models when the rate of income and the probability distribution of the cost of the claim at time t are functions of $Y(t)$, $t \geqslant 0$ (cf. 8).

In this case the risk process

$$X(t) := \int_0^t c(Y(s))\, ds - \int_0^t \int_0^\infty x p(ds, dx), \qquad t \geqslant 0,$$

is again a Markov additive process where $c(y)$, $y \in \mathcal{Y}$, is a finite nonnegative $\mathcal{B}$-measurable function, $p(dt, dx)$ is a point measure on $[0, \infty) \times (0, \infty)$ with the $(P, \mathbb{F}_{\mathcal{G}})$-compensator

$$\Pi([0,t] \times \Gamma) = \int_0^t \pi(Y(s), \Gamma)\, ds, \quad t \geqslant 0, \quad \Gamma \in \mathcal{B}((0, \infty)), \tag{15}$$

such that $\lambda(y) := \pi(y, (0, \infty))$, $y \in \mathcal{Y}$, is a finite nonnegative $\mathcal{B}$-measurable function.

Assume that for some r_∞, $0 < r_\infty \leqslant \infty$, and each $r < r_\infty$, $y \in \mathcal{Y}$

$$h(r, y) := \int_0^\infty (e^{rx} - 1) \pi(y, dx) < \infty. \tag{16}$$

Assume also that for some $\widetilde{\Gamma}_0 \subset [0, r_\infty)$ and each $r \in \widetilde{\Gamma}_0$ there exist (i) a real number $\theta(r)$ and $\tilde{g}_r \in D(A)$, $\tilde{g}_r > 0$, such that

$$A\tilde{g}_r(y) + (h(r,y) - rc(y))\tilde{g}_r(y) = \theta(r)\tilde{g}_r(y), \quad y \in \mathcal{Y}, \tag{17}$$

(ii) a positive real number R, $R \in \widetilde{\Gamma}_0$, called the Lundberg exponent, satisfying

$$\theta(R) = 0, \tag{18}$$

and (iii) for fixed $\underline{t}$, $\bar{t}$, $0 \leqslant \underline{t} < t < \infty$, a positive real number $r_{\underline{t},\bar{t}}$, $r_{\underline{t},\bar{t}} \in \widetilde{\Gamma}_0$, such that

$$\begin{aligned} R_{\underline{t}}^{\bar{t}} &:= \sup_{r \in \tilde{\Gamma}_0} \min(r - \underline{t}\theta(r),\ r - \bar{t}\theta(r)) \\ &= \min(r_{\underline{t},\bar{t}} - \underline{t}\theta(r_{\underline{t},\bar{t}}),\ r_{\underline{t},\bar{t}} - \theta(r_{\underline{t},\bar{t}})). \end{aligned} \tag{19}$$

As before, $R_{\underline{t}}^{\bar{t}}$ is called the finite-time Lundberg exponent.

Proposition 2 (cf. 8, 13). If $E\tilde{g}_R(Y(0)) < \infty$ and assumptions (3)-(5), (15)-(18) are satisfied then for each $u \geqslant 0$

$$\Psi(u) \leqslant \frac{E\tilde{g}_R(Y(0))}{E(\tilde{g}_R(Y(T_u)) \mid T_u < \infty)}\, e^{-Ru}.$$

If $E\tilde{g}_{r_{\underline{t},\bar{t}}}(Y(0)) < \infty$ and assumptions (3)-(5), (15)-(17), (19) are satisfied, then for each $u \geqslant 0$

$$\Psi(u, \bar{t}u) - \Psi(u, \underline{t}u) \leqslant \frac{E\tilde{g}_{r_{\underline{t},\bar{t}}}(Y(0))}{E[\tilde{g}_{r_{\underline{t},\bar{t}}}(Y(T_u)) \mid \underline{t}u < T_u \leqslant \bar{t}u]}\, e^{-R_{\underline{t}}^{\bar{t}} u}.$$

The proof is analogous to the proof of the proposition 1 if we note that under the assumptions of the proposition 2 for each $0 \leqslant s < t$, $r < r_\infty$ P-a.e.

$$E(e^{-r(X(t)-X(s))} \mid \mathcal{F}_s \vee \mathcal{G}) = \exp\left\{ \int_s^t (h(r, Y(v)) - rc(Y(v)))\, dv \right\},$$

and change M_u^r, L^r to $\widetilde{M_u^r}$, $\widetilde{L}^r$, respectively, where

$$\widetilde{M_u^r}(t) = e^{-r(u+X(t))}\tilde{g}_r(Y(t))e^{-\theta(r)t}, \qquad t \geqslant 0,$$

and

$$\widetilde{L}^r(t) = \tilde{g}_r(Y(t)) \exp\left\{ \int_0^t [h(r, Y(v)) - rc(Y(v))]\, dv \right\}, \qquad t \geqslant 0.$$

Obviously, if $c(y) = c_0\lambda(y)+c_1$ and $\pi(y, dx) = \lambda(y)dF(x)$, $y \in \mathcal{Y}$, $c_0 \geqslant 0$, $c_1 \geqslant 0$, then $\theta(r) = \mu(h(r) - c_0 r) - c_1 r$.

REFERENCES

1. H.U. Gerber, Martingales in risk theory, *Mitt. Ver. Schweiz. Vers. Math.*, 73, 205-216, (1973).
2. K. Yosida, *Functional Analysis,* Springer-Verlag, Berlin et al. (1968).
3. J. Jacod, *Calcul stochastique et problémes de martingales.* Lecture Notes in Mth., 714, Springer-Verlag, Berlin et al., (1979).
4. J. Grandell, *Aspects of Risk Theory,* Springer-Verlag, Berlin et al., (1991).
5. P. Embrechts, J. Grandell, and H. Schmidli, Finite-time Lundberg Inequalities in the Cox Case, Royal Institute of Technology, Stockholm, Preprint TRITA-MAT-91-30, (1991).
6. B. Grigelionis, On Markov additive processes. *Liet. Matem. Rinkinys,* (1978), 18(3), 43-47.
7. V.S. Vladimirov, *Equations of Mathematical Physics,* Nauka, Moscow, (1976) (in Russian).
8. S. Asmussen, Risk theory in a markovian environment. *Scand. Actuarial J.*, (1989), 66-100.
9. A. Martin-Löf, Entropy, a useful concept in risk theory. *Scand. Actuarial J.*, (1986), 223-235.
10. T. Björk, J. Grandell, Exponential inequalities for ruin probabilities in the Cox case. *Scand. Actuarial J.*, (1988), 77-111.
11. A. Dassios, P. Embrechts, Martingales and insurance risk. *Commun. Statist.- Stochastic Models,* (1989), 181-217.
12. T. Björk, J. Grandell, Lundberg Inequalities in a Diffusion Environment, Royal Institute of Technology, Stockholm, Working paper, (1991).
13. P. Ney, E. Nummelin, Markov additive processes. I. Eigenvalues and limit theorems. II. Large deviations. *Ann. Probab.*, 15(2), 561-592, 593-609, (1987).

PSEUDO DIFFERENTIAL OPERATORS, FELLER SEMIGROUPS AND THE MARTINGALE PROBLEM

WALTER HOH, NIELS JACOB*
Universität Erlangen-Nürnberg, Mathematisches Institut
Bismarckstraße $1\frac{1}{2}$, D-8520 Erlangen, Germany

In this note we discuss the martingale problem for a large class of pseudo differential operators. There are a lot of recent results concerning the martingale problem for pseudo differential operators and related to this for Lévy type generators, see Komatsu[8,9], Mikulevičius, Pragarauskas[10,11], Negoro, Tsuchiya[12], Stroock[15], Stoock, Varadhan[16] and Tsuchiya[18]. The considerations of this paper are closely related to the thesis[4] of the first named author, see also Hoh[5], and to the second named author's work on Feller semigroups generated by pseudo differential operators. Our work is much inspired by the recent results of J. van Casteren[19].

Acknowledgement: The second named author would like to thank H.-J. Engelbert for his hospitality and support when attending this Winterschool. Financial support for the first named author by the DFG- Forschergruppe "Stochastische Analysis und ihre Anwendungen" is gratefully acknowledged. Special thanks are due to J. van Casteren for stimulating discussions during his visit in Erlangen.

1. FELLER SEMIGROUPS ON $\mathbb{R}^n$

Let us denote by $C_\infty(\mathbb{R}^n)$ the space of all continuous, real-valued func-

*Lecture delivered by the second named author.

tions vanishing at infinity. On $C_\infty(\mathbb{R}^n)$ we have the sup-norm $\|\cdot\|_\infty$ and $(C_\infty(\mathbb{R}^n), \|\cdot\|_\infty)$ is a Banach space with the testfunctions $C_0^\infty(\mathbb{R}^n)$ as a dense subspace.

Definition 1.1. *A* **Feller semigroup** $(T_t)_{t\geq 0}$ *on* $C_\infty(\mathbb{R}^n)$ *is a family of linear operators* $T_t : C_\infty(\mathbb{R}^n) \to C_\infty(\mathbb{R}^n)$ *satisfying*

(i) $T_{s+t} = T_s \circ T_t$, $s, t \geq 0$, *and* $T_0 = \mathrm{id}$ *(semigroup property),*

(ii) $\lim_{t\to 0} \|T_t u - u\|_\infty = 0$ *(strong continuity),*

(iii) $0 \leq T_t u \leq 1$, *whenever* $0 \leq u \leq 1$ *(sub-Markovian property).*

Given a Feller semigroup. Then to a given initial distribution, i.e. a probability measure μ on $\mathbb{R}^n$, there exists a stochastic process $(\Omega, \widetilde{\mathcal{A}}, \widetilde{P}, (\widetilde{X}_t)_{t\geq 0})$ which finite dimensional distributions are given by the transition functions $p_t(x, A) = (T_t\chi_A)(x)$ and μ . Here χ_A denotes the characteristic function of the Borel set A. Moreover, see Ethier, Kurtz[3], p.86, there always exists a càdlàg version of $(\widetilde{X}_t)_{t\geq 0}$. Denoting by $D_{\mathbb{R}^n}$ the space of all functions $\omega : \begin{array}{c} [0,\infty) \to \mathbb{R}^n \\ t \mapsto \omega(t) \end{array}$ which are càdlàg, it follows that for a given initial distribution μ each Feller semigroup determines a unique probability measure P^μ on $(D_{\mathbb{R}^n}, \mathcal{A})$, $\mathcal{A}$ being the σ-field generated by the projections $X_t : D_{\mathbb{R}^n} \to \mathbb{R}^n$, $X_t(\omega) = \omega(t)$, such that $(D_{\mathbb{R}^n}, \mathcal{A}, P^\mu, (X_t)_{t\geq 0})$ is a version of the process $(\Omega, \widetilde{\mathcal{A}}, \widetilde{P}, (\widetilde{X}_t)_{t\geq 0})$. This process we will call the **Feller process** associated with the semigroup $(T_t)_{t\geq 0}$ and the initial distribution μ. Now let again $(T_t)_{t\geq 0}$ be a Feller semigroup on $C_\infty(\mathbb{R}^n)$. Then its **generator** L is defined as the linear operator $L : D(L) \to C_\infty(\mathbb{R}^n)$, where

$$D(L) = \left\{ u \in C_\infty(\mathbb{R}^n),\ \lim_{t\to 0} \frac{1}{t}(T_t u - u) \text{ exists} \right\}$$

and

$$Lu = \lim_{t\to 0} \frac{T_t u - u}{t}.$$

The next theorem gives a characterization of the generators of Feller semigroups:

Theorem 1.1. (Yosida-Hille, see Ethier, Kurtz[3]) *In order that an operator*
$L : D(L) \to C_\infty(\mathbb{R}^n)$ *has a closed extension which generates a Feller semigroup on* $C_\infty(\mathbb{R}^n)$ *the following three assumptions are necessary and sufficient:*

(i) $D(L) \subset C_\infty(\mathbb{R}^n)$ *is dense,*

(ii) L *satisfies the positive maximum principle, i.e. for all* $u \in D(L)$ *it follows that* $\sup_{x\in\mathbb{R}^n} u(x) = u(x_0) \geq 0$ *implies* $Lu(x_0) \leq 0$,

(iii) for some $\lambda \geq 0$ *the range of* $\lambda - L$ *is dense in* $C_\infty(\mathbb{R}^n)$.

By a result of Ph. Courrège[2], p. 2-37, we find

Theorem 1.2. *An operator* $L : C_0^\infty(\mathbb{R}^n) \to C(\mathbb{R}^n)$ *satisfies the positive maximum principle on* $C_0^\infty(\mathbb{R}^n)$ *if there exists a continuous function* $L : \mathbb{R}^n \times \mathbb{R}^n \to \mathbb{C}$ *such that for each* $x \in \mathbb{R}^n$ *the function* $L(x,\cdot) : \mathbb{R}^n \to \mathbb{C}$ *is negative definite and* L *is given by*

$$-Lu(x) = -L(x,D)u(x) = (2\pi)^{-n/2} \int_{\mathbb{R}^n} e^{ix\cdot\xi}(-L(x,\xi))\hat{u}(\xi)\,d\xi, \quad (1.1)$$

where $\hat{u}$ *denotes the Fourier transform of* u.

Recall that a function $a : \mathbb{R}^n \to \mathbb{C}$ is negative definite if and only if for any $m \in \mathbb{N}$ and any collection $\xi^1,\dots,\xi^m \in \mathbb{R}^n$ the matrix $(a(\xi^i) + \overline{a(\xi^j)} - a(\xi^i - \xi^j))_{i,j=1,\dots,m}$ is positive Hermitian. We refer to Berg, Forst[1] where a lot of properties and examples of negative definite functions are given. We restrict ourselves to real-valued continuous negative definite functions. In this case one knows that $0 \leq L(x,\xi) \leq \tilde{c}(1 + |\xi|^2)$ holds, thus the operator $L(x,D)$ is well defined on $C_0^\infty(\mathbb{R}^n)$. We call this operator the **pseudo differential operator** generated by the **symbol** $L(x,\xi)$. Note that in general $L(x,\xi)$ does not belong to a classical symbol class, see Hörmander[6], Oleinik, Radkevic[13] or Taylor[17], since a continuous negative definite function need not be differentiable.For this reason one cannot use some standard symbolic calculus in order to handle these operators. The next result is proved in Jacob[7].

Theorem 1.3. *Let* $L : \mathbb{R}^n \times \mathbb{R}^n \to \mathbb{R}$ *be a continuous function such that for each* $x \in \mathbb{R}^n$ *the function* $L(x,\cdot) : \mathbb{R}^n \to \mathbb{R}$ *is negative definite. Further let* $x_0 \in \mathbb{R}^n$ *be fixed and assume that* $L(x,\xi)$ *has the decomposition*

$$L(x,\xi) = L(x_0,\xi) + (L(x,\xi) - L(x_0,\xi)) = L_1(\xi) + L_2(x,\xi)$$

and that the following assumptions are fulfilled:

(i) There exists a continuous negative definite function $a^2 : \mathbb{R}^n \to \mathbb{R}$ *such that* $L_1(\xi) \le \gamma_1(1 + a^2(\xi))$ *for all* $\xi \in \mathbb{R}^n$, *and* $a^2(\xi) \ge c\ |\xi|^r$ *for some* $r > 0$ *and all* $\xi \in \mathbb{R}^n$, $|\xi| \ge \varrho \ge 0$.

(ii) For $q \in \mathbb{N}$ *sufficiently large* ($q \ge n + 3\left[\frac{n+1}{r}\right] + 6$ *is sufficient) we suppose that* $L_2(\cdot, \xi) \in C^q(\mathbb{R}^n)$ *and that* $|\partial_x^\alpha L_2(x,\xi)| \le \varphi_\alpha(x)(1 + a^2(\xi))$ *holds for all* $\alpha \in \mathbb{N}_0^n$, $|\alpha| \le q$, *with* $\varphi_\alpha \in L^1(\mathbb{R}^n)$.

(iii) For some $\gamma_0 > 0$ *it is supposed that* $L_1(\xi) \ge \gamma_0 a^2(\xi)$ *holds for all* $\xi \in \mathbb{R}^n$, $|\xi| \ge \varrho \ge 0$.

(iv) Finally we assume that

$$\sum_{|\alpha| \le q} \|\varphi_\alpha\|_{L^1} \cdot \int_{\mathbb{R}^n} (1 + |\tau|^2)^{(1-q)/2}\, d\tau \le d \cdot \gamma_0, \tag{1.2}$$

where d is an explicitly known (small) constant.

Then $-L(x,D)$ *has a closed extension generating a Feller semigroup on* $C_\infty(\mathbb{R}^n)$.

Remark 1.1. *The condition* (1.2) *should be regarded as a restriction to the oscillation of the coefficients of the operator* $L(x,D)$.

The proof of Theorem 1.3 requires hard analysis for the operator $L(x,D)$ in a scala of anisotropic Sobolev spaces defined by

$$H^{a^2,s}(\mathbb{R}^n) = \left\{u \in \mathcal{S}'(\mathbb{R}^n), (1 + a^2(\cdot))^s \hat{u}(\cdot) \in L^2(\mathbb{R}^n)\right\}.$$

In particular one needs a Gårding inequality, certain commutator estimates and regularity results for weak solutions of the equation $L(x,D)u + \lambda u = f$, $\lambda \in \mathbb{R}$.

In the next section we will try to weaken some of the conditions of Theorem 1.3, especially assumption (iv).

2. FELLER SEMIGROUPS AND THE MARTINGALE PROBLEM

In order to get a Markov process by starting with a pseudo differential operator one can use different methods. Besides constructing a family of transition functions using the semigroup generated by the operator there are further possibilities. For example one can try to prove that the martingale problem is uniquely solvable.

Definition 2.1. *Let $(\Omega, \mathcal{A}, P)$ be a probability space and $(\Omega, \mathcal{A}, P, (X_t)_{t\geq 0})$ a stochastic process with state space $\mathbb{R}^n$. Further let $L : D(L) \to \mathbb{R}^n$ be a linear operator, $C_0^\infty(\mathbb{R}^n)$ being dense in $D(L)$, and let μ be a probability measure on $\mathbb{R}^n$. In addition suppose that $(X_t)_{t>0}$ is progressively measurable with respect to the canonical filtration $(\mathcal{F}_t^0)_{t\geq 0}$. We say that $(X_t)_{t\geq 0}$ is a solution of the* **martingale problem** *with respect to $-L$ and μ, if and only if*

(i) $P_{X_0} = \mu$, where P_{X_0} denotes the distribution of X_0,

(ii) for all $\varphi \in C_0^\infty(\mathbb{R}^n)$

$$M_t^\varphi := \varphi(X_t) - \int_0^t (-L\varphi)(X_s)\, ds, \ t \geq 0, \tag{2.1}$$

is a P-martingale with respect to $(\mathcal{F}_t^0)_{t\geq 0}$.

In the case where we are interested in, namely solving the martingale problem for a certain class of pseudo differential operators, we can assume that the process solving the martingale problem is defined on the measure space $(D_{\mathbb{R}^n}, \mathcal{A})$, where $\mathcal{A}$ is the σ-field generated by the projections $X_t : D_{\mathbb{R}^n} \to \mathbb{R}^n$, $X_t(\omega) = \omega(t)$, and the family $(X_t)_{t\geq 0}$ itself gives the process, see Ethier, Kurtz[3], chap. 4.3. Thus the only thing to be determined by L and μ is the probability measure P. We call this reduced problem the $D_{\mathbb{R}^n}$-martingale problem for the operator $-L$ and the initial distribution μ. Furthermore, we say that the $D_{\mathbb{R}^n}$-martingale problem is **well-posed**, if for any initial distribution μ there exists a unique probability measure P^μ on $(D_{\mathbb{R}^n}, \mathcal{A})$ such that $(D_{\mathbb{R}^n}, \mathcal{A}, P^\mu, (X_t)_{t\geq 0})$ solves the $D_{\mathbb{R}^n}$-martingale problem.

It is easy to see that, if a pseudo differential operator L has an extension generating a conservative Feller semigroup, then the $D_{\mathbb{R}^n}$-martingale problem is well-posed. Thus for the operators considered in Theorem 1.3 the $D_{\mathbb{R}^n}$-martingale problem is well-posed. Next we want to weaken the assumptions on these operators and still prove the well-posedness of the $D_{\mathbb{R}^n}$-martingale problem for a larger class of pseudo differential operators. For this we will use a certain localization procedure. Let $U \subset \mathbb{R}^n$ be an open set and τ_U be defined by

$$\tau_U = \inf\{t \geq 0, X_t \notin U \text{ or } (t > 0 \text{ and } X_{t-} \notin U)\}, \tag{2.2}$$

where $(D_{\mathbb{R}^n}, \mathcal{A}, P^\mu, X_t)_{t\geq 0})$ is a càdlàg process. Then τ_U is a $(\mathcal{F}_t^0)$-stopping time. Further we give

Definition 2.2. *Let L be a pseudo differential operator as before, $U \subset \mathbb{R}^n$ an open set and μ a probability measure on $\mathbb{R}^n$. Then a probability measure $\widetilde{P}^\mu$ on $D_{\mathbb{R}^n}$ is called a solution of the* **stopped martingale problem** *for $-L$, μ and U, if and only if*

(i) $\widetilde{P}^\mu_{X_0} = \mu$,

(ii) $\widetilde{P}^\mu\{X_t = X_{t\wedge\tau_U}$ *for all* $t \geq 0\} = 1$,

(iii) $\varphi(X_{t\wedge\tau_U}) - \int_0^{t\wedge\tau_U}(-L\varphi)(X_s)\,ds,\ t \geq 0$, *is a* $(\mathcal{F}_t^0)_{t\geq 0}$*-martingale with respect to* $\widetilde{P}^\mu$ *and for all* $\varphi \in C_0^\infty(\mathbb{R}^n)$.

First we note

Theorem 2.1. *Suppose that $L(x,\xi)$ satisfies condition (i) and (ii) of Theorem 1.3 and that for any initial distribution μ there exists a solution of the $D_{\mathbb{R}^n}$-martingale problem with respect to $-L$ and μ. Further suppose that there is a covering of $\mathbb{R}^n$ by open sets U_k, $k \in \mathbb{N}$, such that for each $k \in \mathbb{N}$ there exists a symbol $L^k(x,\xi)$ fulfilling the assumptions of Theorem 1.3. In addition let us assume that $L^k(x,\xi) = L(x,\xi)$ holds for all $x \in U_k$ and all $\xi \in \mathbb{R}^n$. Then the $D_{\mathbb{R}^n}$-martingale problem for $-L^k(x,D)$ as well as the $D_{\mathbb{R}^n}$-martingale problem for $-L(x,D)$ is well-posed.*

The proof of this theorem uses first the fact that the $D_{\mathbb{R}^n}$-martingale problem for each of the operators $-L^k(x,D)$ is well-posed since they do extend to a generator of a Feller semigroup. Then it is shown that the stopped $D_{\mathbb{R}^n}$-martingale problem for each $-L^k(x,D)$ and U_k is also well-posed. This however implies that the $D_{\mathbb{R}^n}$-martingale problem for $-L(x,D)$ is well-posed. Details for these results may be found in Ethier, Kurtz[3], chap.4.6. A rather detailed calculation for a special class of pseudo differential operators is given in Hoh[4] and the theorem could be proved along this lines.

The next result is taken from Hoh[4]:

Theorem 2.2. *Let $L(x,D) = \sum_{j=1}^{N} b_j(x)\,a_j^2(D)$ be a pseudo differential operator satisfying the following conditions:*

(i) $a_j^2 : \mathbb{R}^n \to \mathbb{R}$ *is a continuous negative definite function and* $a_j^2(0) = 0$,

(ii) $b_j : \mathbb{R}^n \to \mathbb{R}$ *is a non-negative, continuous and bounded function,*

(iii) there exist $r > 0$ and $c > 0$ such that

$$1 + \sum_{j=1}^{N} a_j^2(\xi) \geq c \cdot (1 + |\xi|^2)^{r/2} \text{ holds.}$$

Let m_0 be the smallest integer such that $r \cdot m_0 > \frac{n}{2} + 2$. If $b_j(x) > 0$ for all $x \in \mathbb{R}^n$ and $b_j \in C^{2(m_0+1)+n}(\mathbb{R}^n)$, then the $D_{\mathbb{R}^n}$-martingale problem for $-L$ is well-posed.

The proof of this theorem goes in two steps. First it is proved that the $D_{\mathbb{R}^n}$-martingale problem for $-L$ is well-posed if the functions b_j satisfy an additional oscillation condition and then a partition of unity argument is used.

Combining Theorem 1.3 and Theorem 2.2 to ideas of the proof of Theorem 2.2 leads us to the following **conjecture**:

Let $L : \mathbb{R}^n \times \mathbb{R}^n \to \mathbb{R}$ be a continuous function such that for each $x \in \mathbb{R}^n$ the function $L(x, \cdot) : \mathbb{R}^n \to \mathbb{R}$ is negative definite and $L(x, 0) = 0$. Further let $a^2 : \mathbb{R}^n \to \mathbb{R}$ be a fixed continuous negative definite function satisfying with some $r > 0$ the inequality $a^2(\xi) > c\,|\xi|^r$ for all $\xi \in \mathbb{R}^n$, $|\xi| \geq \varrho \geq 0$, and for simplicity suppose that in the Lévy-Khinchin representation of a^2 the quadratic form is identically zero. Moreover let us suppose

(i) with $q \in \mathbb{N}$ sufficiently large assume $L(\cdot, \xi) \in C^q(\mathbb{R}^n)$ for all $\xi \in \mathbb{R}^n$ and

(ii) there is a continuous function $c : \mathbb{R}^n \to \mathbb{R}$, $c(x) > 0$ for all $x \in \mathbb{R}^n$ such that

$$L(x, \xi) \geq c(x) a^2(\xi) \tag{2.3}$$

holds for all $x \in \mathbb{R}^n$ and all $\xi \in \mathbb{R}^n$, $|\xi| \geq \varrho \geq 0$.

Then the $D_{\mathbb{R}^n}$-martingale problem for $-L(x, D)$ is well-posed.

Unfortunately until now we can not prove this conjecture in full generality, but only in situations where one knows a posteriori some additional properties of certain derivatives of $L(x, \xi)$ with respect to x.

Suppose that we know the well-posedness of the $D_{\mathbb{R}^n}$-martingale problem for a certain pseudo differential operator, see for example Theorem 2.2. Then a natural question is, whether the operator $-L(x, D)$ itself extends to a generator of a Feller semigroup. The next result is due to J. van Casteren[19], Proposition 2.6, and gives a certain answer to this question.

Theorem 2.3. *Let L be a densely defined operator for which the martingale problem is well-posed. Then there exists a unique closed linear extension L_0 of L, which is the generator of a Feller semigroup.*

REFERENCES

1. C. Berg, G. Forst, Potential theory on locally compact Abelian groups. Ergebnisse der Mathematik und ihrer Grenzgebiete, **II.** Ser. Bd.**87**, Springer Verlag, Berlin - Heidelberg - New York (1975)

2. Ph. Courrége, Sur la forme intégro-différentielle des opérateurs de C_k^∞ dans C satisfaisant au principe du maximum. Sém. Théorie du Potentiel (1965/66) 38 p.

3. S.N. Ethier, Th.G. Kurtz, Markov processes — characterization and convergence. Wiley Series in Probability and Mathematical Statistics, John Wiley, New York - Chicester - Brisbane - Toronto - Singapore (1986)

4. W. Hoh, Das Martingalproblem für eine Klasse von Pseudodifferentialoperatoren. Dissertation, Universität Erlangen - Nürnberg, Germany (1992)

5. W. Hoh, On the martingale problem for a class of pseudo differential operators. (in preparation)

6. Hörmander, L. The analysis of linear partial differential operators III, Grundlehren der mathematischen Wissenschaften, Vol. **274**. Berlin - Heidelberg - New York - Tokyo: Springer Verlag (1985)

7. N. Jacob, A class of Feller semigroups generated by pseudo differential operators (submitted)

8. T. Komatsu, On the martingale problem for generators of stable processes with perturbations. Osaka J.Math. **21**, 113-132 (1984)

9. T. Komatsu, Pseudo-differential operators and Markov processes. J.Math.Soc. Japan **36**, No.**3**, 387-418 (1984)

10. R. Mikulevičius, H. Pragarauskas, On the existence and uniqueness of solutions to the martingale problem. Preprint, Vilnius, 1-62 (1990)

11. R. Mikulevičius, H. Pragarauskas, On the existence and uniqueness of solutions to the martingale problem associated with Lipschitz continuous Lévy's generator. Preprint, Vilnius 1-11

12. A. Negoro, M. Tsuchiya, Stochastic processes and semigroups associated with degenerate Lévy generating operators. Stochastics and Stoch.Rep. **26**, 29-61 (1989)

13. O.A. Oleinik, E.V. Radkevic, Second order equations with non-negative characteristic form. Providence R.I.: Amer.Math.Soc., New York - London: Plenum Press (1973)

14. D. Revuz, M. Yor, Continuous martingales and Brownian motion. Grundlehren der mathematischen Wissenschaften, Vol. **293**. Berlin: Springer Verlag (1991)

15. D.W. Stroock, Diffusion processes associated with Lévy generators. Z. Wahr. verw. Geb. **32**, 209-244 (1975)

16. D.W. Stroock, S.R.S. Varadhan, Multidimensional diffusion processes. Grundlehren der mathematischen Wissenschaften **233**, Springer Verlag, Berlin - Heidelberg - New York (1979).

17. M. Taylor, Pseudodifferential operators, Princeton Math. Ser., Vol. **34**. Princeton: Princeton University Press (1981)

18. M. Tsuchiya, Lévy measure with generalized polar decomposition and the associated SDE with jumps. Stochastics and Stoch.Rep. **38**, 95-117 (1992)

19. J.A. van Casteren, On martingales and Feller semigroups. Results in Mathematics **21**, 274-288 (1992)

THE PRICING OF LOOK BACK OPTIONS AND A FUBINI THEOREM FOR ITÔ- AND LEBESGUE-INTEGRALS

RALF KORN
Fachbereich Mathematik, Johannes Gutenberg-Universität Mainz, Postfach 3980, Saarstraße 21, 6500 Mainz

Abstract The pricing of look back options in the usual Black & Scholes model is discussed. Two approaches to get a representation of the payment of a look back option are leading to a Fubini-theorem.

1. SOME BASIC REMARKS

Consider a simple financial market model with constant coefficients consisting of a stock and a bond with price dynamics given by

$$dS_t = \mu\, S_t\, dt + \sigma\, S_t\, d\,W_t \quad , \quad S_0 = s \qquad \text{“stock”} \tag{1.1}$$

$$dB_t = r\, B_t\, dt \quad , \quad B_0 = 1 \qquad \text{“bond”} \tag{1.2}$$

where $\mu, r, \sigma \in I\!R, \sigma \neq 0, s > 0, t \in [0,T], T \in I\!R$ fixed, $\{W_t\}_{t\in[0,T]}$ a standard one dimensional Brownian motion on a complete probability space (Ω, f, P) which is equipped with the P-augmentation $\{f_t\}_{t\in[0,T]}$ of the natural Brownian filtration.

Definition 1
A European option with maturity T consists of a non-negative f_T-measurable payment B at the terminal time T with

$$E\,B^{\mu} < \infty \quad \textit{for some } \mu > 1. \tag{1.3}$$

For deriving the fair price of a European option by the principles of duplication and no arbitrage we need some definitions:

Definition 2

1. *A trading strategy is a pair (ψ, φ) of predictable processes with*

$$\int_0^T \psi_s B_s \, ds \; < \; \infty \quad a.s. \tag{1.4}$$

$$\int_0^T (\varphi_s \sigma S_s)^2 \, ds \; < \; \infty \quad a.s. \tag{1.5}$$

$x := \psi_0 B_0 + \varphi_0 S_0$ *is called the initial wealth of the trading strategy.*

2. $$X_t := \varphi_t S_t + \psi_t B_t, \quad t \in [0, T] \tag{1.6}$$

is called the wealth process of an investor who follows the trading strategy (ψ, φ).

3. *A trading strategy (ψ, φ) is called self-financing if*

$$X_t = x + \int_0^t \varphi_s \, dS_s + \int_0^t \psi_s \, dB_s \; \forall \, t \in [0, T] \; a.s. \tag{1.7}$$

holds (that means that all changes in the wealth of an investor who follows a self-financing strategy arise from gains/losses of investment).

4. *A self-financing trading strategy (ψ, φ) is said to generate the European Option with payment B (or a generating strategy) if*

$$B = \psi_T B_T + \varphi_T \, S_T \quad a.s. \tag{1.8}$$

Now the *fair price* p of a European option is defined (as usual) as

$$p \;=\; \inf\{x > 0 : \; \exists \text{ a generating strategy } (\psi, \varphi) \text{ with initial wealth } x\}. \tag{1.9}$$

It can be shown that this is the only price in our market model that prevents arbitrage opportunities. Let $\tilde{P}$ be the unique probability measure which is equivalent to P on f_T such that S_t/B_t is a $\tilde{P}$-martingale ($\tilde{P}$ is often called the martingale measure (compare for example Harrison & Pliska[3]). It is now easy to compute the fair price p.

Theorem 1
Let B be the payment of a European option. We have

$$p = \tilde{E}(Be^{-rT}) \tag{1.10}$$

(where $\tilde{E}$ means expectation with respect to $\tilde{P}$) and there exists a unique (up to equivalence) generating strategy (ψ, φ) with initial wealth p and corresponding wealth process X_t given by

$$X_t = \tilde{E}(B\, e^{-r(T-t)}/f_t) \qquad a.s. \tag{1.11}$$

Proof. See Harrison & Pliska[3] in combination with Karatzas & Shreve[6], page 378.

Example
The most famous example is the European call option which is given by

$$B = (S_T - c)^+, \tag{1.12}$$

where c is a positive constant. In this case we can compute all relevant objects explictly:

$$\begin{aligned} X_t &= \underbrace{S_t\,\Phi\left(\frac{1}{\sigma\sqrt{T-t}}\left(ln(\frac{S_t}{c}) + (r+\frac{1}{2}\sigma^2)(T-t)\right)\right)}_{=:\varphi_t} \qquad (1.13)\\ &\quad -e^{rt}\cdot \underbrace{e^{-rT}c\Phi\left(\frac{1}{\sigma\sqrt{T-t}}\left(ln(\frac{S_t}{c}) + (r-\frac{1}{2}\sigma^2)(T-t)\right)\right)}_{=:-\psi_t} \\ X_T &= B \quad a.s. \qquad p = X_0 \qquad (1.14) \end{aligned}$$

(1.13) and (1.14) constitute the famous Black & Scholes formula (compare Black & Scholes[1], Karatzas & Shreve[6], page 378).
Φ is the distribution function of the standard normal distribution.

2. THE PRICING OF LOOK BACK OPTIONS

Definition 3
A look back option is a European option with

$$B = F(\tilde{W}) \tag{2.1}$$

where $F : C[0,T] \to I\!R_+$ *is a functional (with (1.3)),* $\tilde{W}$ *a one dimensional standard* $\tilde{P}$*-Brownian motion constructed from* W *via Girsanov's theorem, (see Karatzas & Shreve*[6]*), i.e.*

$$\tilde{W}_t = W_t + \frac{b-r}{\sigma}t, \quad t \in [0,T]. \tag{2.2}$$

Remarks

1. *The form of the payment* B *being a functional of* $\tilde{W}$ *seems to be unnatural because one would have expected a functional of* W*. We define it in such a way because this form is easier to handle in our computations. Notice further that relation (2.2) allows us to transform a functional* $F^*(W)$ *in a functional* $F(\tilde{W})$*.*

2. *In contrast to a European call where the payment only depends on the stock price at the terminal time* T *the payment of a look back option depends on the (whole) path of the stock price in [0,T]. For ease of exposition we assume from now on*

$$r = 0. \tag{2.3}$$

The fair price p^F *of a look back option given by equation (2.1) is easily found by applying Theorem 1 as*

$$p^F = \tilde{E}F(\tilde{W}). \tag{2.4}$$

The computation of the corresponding generating strategy is more involved. It is demonstrated by the "look back version" of the European call:

Example
Let

$$\begin{aligned} B = F(\tilde{W}) &= \int_0^T (se^{\sigma\tilde{W}_t - \frac{1}{2}\sigma^2 t} - c)^+ dt \qquad (2.5) \\ &= \int_0^T (S_t - c)^+ dt. \end{aligned}$$

where c is a positive constant. From equations (1.13) and (1.14) we obtain (remember $r = 0$, $dB_s = 0$)

$$(S_t - c)^+ = \tilde{E}(S_t - c)^+ + \int_0^t \Phi\left(\frac{1}{\sigma\sqrt{t-s}}\left(ln(\frac{S_s}{c}) + \frac{1}{2}\sigma^2(t-s)\right)\right) dS_s, \tag{2.6}$$

hence

$$F(\tilde{W})$$

$$= \int_0^T \left[\tilde{E}(S_t - c)^+ + \int_0^t \Phi\left(\frac{1}{\sigma\sqrt{t-s}}\left(ln(\frac{S_s}{c}) + \frac{1}{2}\sigma^2(t-s)\right)\right) dS_s\right] dt$$

$$= p^F + \int_0^T \int_0^t \Phi\left(\frac{1}{\sigma\sqrt{t-s}}\left(ln(\frac{S_s}{c}) + \frac{1}{2}\sigma^2(t-s)\right)\right) dS_s\, dt. \tag{2.7}$$

So if we could interchange the order of integration in the double integral we could identify φ_t (and ψ_t via (1.6)). Before that we do another approach (compare Föllmer[2]).
The basic tool of this approach is

Theorem 2 ("Clark's formula")
Let (Ω, f, P) be the probability space with $\Omega = C[0,T]$, P Wiener measure, $\{W_t\}_{t\in[0,T]}$ the coordinate process (i.e. a Brownian motion), $\{f_t\}_{t\in[0,T]}$ the P-augmentation of $f_t^W := \sigma(W_u, u \le t)$, $f := \sigma(\cup_{t\in[0,T]} f_t)$. If the functional $F : C[0,T] \to I\!R$ fulfills

$$E(F(W)^2) < \infty \tag{2.8i}$$

$\exists$ a constant γ such that we have for every $\varphi \in \Omega$

$$|F(W+\varphi) - F(W)| \le \gamma\|\varphi\|_\infty \tag{2.8ii}$$

$\exists$ a kernel F' from Ω to $[0,T]$ such that for all $\varphi \in C^1[0,T]$ we have

$$\lim_{\varepsilon\to 0} \frac{F(W+\varepsilon\varphi) - F(W)}{\varepsilon} = \int_0^T F'(W, dt)\varphi_t \quad a.s. \tag{2.8iii}$$

then

$$F(W) = EF(W) + \int_0^T E(F'(W,(t,T]/f_t)dW_t \quad a.s. \tag{2.9}$$

Proof. See Rogers & Williams[7], page 90-92.

Remarks

3. *All assumptions (2.8) are fulfilled if F is Fréchet differentiable with bounded derivative.*

4. *Unfortunately, the functional F of our example (the "look back version" of the European call) does not satisfy assumption (2.8ii) but careful inspection of the proof of Clark's formula given by Rogers & Williams shows that one can weaken condition (2.8ii) in such a way that our case is covered.*

5. *Note that we have*
$$dS_t = \sigma\, S_t\, d\tilde{W}_t \tag{2.10}$$
so we get the generating strategy φ from equation (2.9) as
$$\varphi_t = \frac{E(F'(W,(t,T])/f_t)}{\sigma\, S_t}. \tag{2.11}$$
(and ψ from equation (1.6) together with equation (2.11)).

Example (continued)

Application of Theorem 2 (together with Remark 4) and some computations are leading to the following (second) representation of $F(\tilde{W})$

$$F(\tilde{W}) = \nu^F + \int_0^T \underbrace{\int_t^T \Phi\left(\frac{1}{\sigma\sqrt{s-t}}\left(ln(\frac{S_t}{c}) + \frac{1}{2}\sigma^2(s-t)\right)\right) ds}_{=:\varphi_t}\, dS_t \tag{2.12}$$

and we can identify φ_t.
On the other hand, the two representations (2.7) and (2.12) result in the following equality for the two double integrals

$$\int_0^T \int_0^t \Phi\left(\frac{1}{\sigma\sqrt{t-s}}\left(ln(\frac{S_s}{c}) + \frac{1}{2}\sigma^2(t-s)\right)\right) dS_s dt$$
$$= \int_0^T \int_t^T \Phi\left(\frac{1}{\sigma\sqrt{s-t}}\left(ln(\frac{S_t}{c}) + \frac{1}{2}\sigma^2(s-t)\right)\right) ds dS_t. \tag{2.13}$$

3. LOOK BACK OPTIONS AND A FUBINI THEOREM

A "kind of generalization" of equation (2.13) is

Theorem 3

Let $f : [0,T] \times \mathbb{R} \to \mathbb{R}$ be continuous and continuously differentiable with respect to the second variable, $\int_0^T Ef(t,W_t)dt < \infty$, $E\left(\left(\int_0^T f(t,W(t))dt\right)^2\right) < \infty$, where $\{W_t\}_{t\in[0,T]}$ is a one-dimensional Brownian motion.
Let $f'(t,W_t)$ be $P\otimes\lambda$ integrable, where λ is Lebesgue-measure on $[0,T]$, $f'(t,x) := \frac{\partial}{\partial x} f(t,x)$.
Then we have:

$$\int_0^T \int_t^T E(f'(s,W_s)/f_t)dsdW_t = \int_0^T \int_0^t E(f'(t,W_t)/f_s)dW_s dt \quad a.s. \tag{3.1}$$

$$\int_0^T \int_t^T E(f'(s,W_s)/f_t)dsdW_t = \int_0^T [f(t,W_t) - Ef(t,W_t)]dt \quad a.s. \tag{3.2}$$

Remark

1. *Equation (3.2) says that we can write the stochastic integral on the left hand side as a Lebesgue integral of the centralized process $f(t,W_t) - Ef(t,W_t)$.*

Proof of Theorem 3. Notice that if $h(s,W_s)$ is $P\otimes\lambda$-integrable we have

$$E\left(\int_t^T h(s,W_s)/f_t\right) = \int_t^T E(h(s,W_s)/f_t)ds \quad a.s. \tag{3.3}$$

Define

$$F(W) := \int_0^T f(t,W_t)dt. \tag{3.4}$$

Under the assumptions of Theorem 3 we get

$$F'(W,(t,T]) = \int_t^T f'(s,W_s)ds \ a.s. \tag{3.5}$$

$$F(W) \overset{(2.9)}{=} EF(W) + \int_0^T E(F'(W,(t,T])/f_t)dW_t \tag{3.6}$$

$$\overset{(3.3)}{=} EF(W) + \int_0^T \int_t^T E(f'(s,W_s)/f_t)ds\, dW_t. \quad a.s.$$

Define

$$F^t(W) := f(t, W_t). \tag{3.7}$$

Then we have

$$F^t(W) = f(t, W_t) \overset{(2.9)}{=} Ef(t, W_t) + \int_0^t E(f'(t, W_t)/f_s)dW_s \ a.s. \tag{3.8}$$

$$\begin{aligned} F(W) &= \int_0^T f(t, W_t)dt \\ &\overset{(3.8)}{=} EF(W) + \int_0^T \int_0^t E(f'(t, W_t)/f_s)dW_s \, dt \quad a.s. \end{aligned} \tag{3.9}$$

Now the equations (3.6) and (3.9) imply equation (3.1) and equation (3.2) follows from equation (3.1) and

$$f(t, W_t) - Ef(t, W_t) = \int_0^t E(f'(t, W_t)/f_s)dW_s \quad a.s. \qquad \Box \tag{3.10}$$

Remarks

2. *Theorem 3 can be formulated for a wider class of functionals (compare the conditions for Clark's formula) but for ease of exposition we restrict ourselves to the case covered by the assumptions of Theorem 3.*

3. *There are related (more general) theorems of the Fubini type for Itô integrals by Kallianpur & Striebel*[4] *proved by L^2-methods. One reason for presenting Theorem 3 above is that its proof follows directly from doing two different approaches to the computation of the generating strategy of a look back option, so the theorem can be viewed as a "corollary to look back pricing."*

4. *A consequence of Theorem 3 is that we can get the generating strategy of a look back option with a terminal payment*

$$F(\tilde{W}) = \int_0^T f(t, \tilde{W}_t) \, dt \tag{3.11}$$

directly from options with maturity t and terminal payment

$$F^t(\tilde{W}) = f(t, \tilde{W}_t) \tag{3.12}$$

for all $t \in [0, T]$ by computing their generating strategies with the usual methods (compare Karatzas[5]*), integrating them and then interchanging the order of integration in the sense of (3.1).*

References

1 F. Black, M. Scholes, The pricing of options and corporate liabilities. Journal of Political Economy **81**, 637-659, 1973.

2 H. Föllmer, Probabilistic Aspects of Options. *Preprint*, 1991.

3 J.M. Harrison, S.R. Pliska, Martingales and stochastic integrals in the theory of continuous trading. Stochastic Processes and Applications **11**, 215-260.

4 G. Kallianpur, C. Striebel, Stochastic differential equations occurring in the estimation of continuous parameter stochastic processes, Theory of Probability and its Applications **14**, 567-594, 1969.

5 I. Karatzas, Optimization problems in the theory of continous trading, SIAM Journal on Control and Optimization **27**, 1221-1259, 1989.

6 I. Karatzas, S.E. Shreve, Brownian Motion and Stochastic Calculus, New York: Springer, 1987.

7 L.C.G. Rogers, D. Williams, Diffusions, Markov Processes and Martingales Volume II. Chichester: John Wiley and Sons, 1987.

LINEAR LOCAL OPERATORS AND MULTIDIMENSIONAL CONTINUOUS STRONG MARKOV PROCESSES WITH TRANSITION DENSITIES

JÖRG-UWE LÖBUS
Mathematische Fakultät, Friedrich-Schiller-Universität Jena, O-6900 Jena, Germany

1. INTRODUCTION

Linear local operators have been introduced by E.B. Dynkin (cf. [5], V.4) in order to define the characteristic operator of a Markov process. Since that time numerous authors - from various points of view - have discussed the relations between such operators and Markov processes.

In this note we consider a certain class $\mathfrak{A}$ of linear local operators which satisfy

(i) the local maximum principle

and have, on any regular domain,

(ii) at least one function belonging to the domain of definition the image of which is strictly positive

and

(iii) solutions of the Dirichlet problem for a dense set of continuous boundary functions.

(The potential theoretical notions as regular domains, local maximum principle etc. will be explained in section 2.)

The axioms (i)-(iii) are introduced with the

following aim: We want to find representations for operators belonging to the class $\mathfrak{A}$. The basic one is given in Theorem 2.1. It asserts that an operator $A\in\mathfrak{A}$ can be represented by

$$Af(x)=\lim_{V\downarrow x}\frac{\int_{\partial V} f(y)\,\mu_{x,V}(dy)-f(x)}{\frac{1}{Ag(x)}\left(\int_{\partial V} g(y)\,\mu_{x,V}(dy)-g(x)\right)} \tag{1}$$

whrer g is an arbitrary strictly subharmonic function and $\mu_{x,V}$ is the harmonic measure with respect to the regular domain V and the point $x\in V$.

Furthermore, under some additional conditions, we give representations in the form of potentials (Theorem 3.1(a) and Theorem 5.1),

$$f=-\int K(\cdot,y)\,Af(y)\,M(dy), \tag{2}$$

and in the form of differential operators with generalized coefficients (Theorem 3.1(b)),

$$\int fB^*\phi\,dx=\int \phi Af\,dM,\ \phi\in C_0^\infty. \tag{3}$$

In order to obtain the last representation we assume that the operator A has harmonic functions of a second order elliptic differential operator B. (We regard B as a linear local operator belonging to the class $\mathfrak{A}$).

Let us return to the equation (1). Assuming Ag≡1 the denominator of the right hand side of the equation (1) is the solution of the equation $A\tilde{g}=-1$ on V satisfying the boundary condition $\tilde{g}|_{\partial V}=0$. For this reason, the representation (1) has the form of a characteristic operator $\mathcal{A}$ of a continuous strong Markov process $X=(X_t,\zeta,\mathcal{M}_t,P_x)$,

$$\mathcal{A}f(x)=\lim_{V\downarrow x}\frac{\int_{\partial V} f(y)\ P_x(X_{\tau(V)}\in dy)-f(x)}{E_x\tau(V)},$$

(E_x denotes the mathematical expectation with respect to the measure P_x and $\tau(V)$ is the first exit time from the domain V).

Now, one can ask: Does there exist a continuous strong Markov process the characteristic operator of which is a given linear local operator belonging to the class $\mathfrak{A}$? The answer is "yes", see Theorem 2.2. In particular, we show that (under an additional condition) such processes admit *transition densities*.

In the one-dimensional case the generator of a conservative regular continuous strong Markov process can be represented by

$$Af=D_m D_x^+ f,\ f\in D_A. \tag{4}$$

Endowed with suitable boundary conditions, this operator is selfadjoint in $L^2(m)$ (m is the measure arising from the equation (4)). Now, arguments from spectal theory show that the functions

$$p(t,x,.):=\int_{-\infty}^{0} e^{\lambda t}\phi_1(x,\lambda)\phi_2(.,\lambda)\ \tau(d\lambda)$$

are the transition densities (with respect to the measure m) of the process considered. Here τ is the associated spectral measure and ϕ_1 and ϕ_2 are suitable solutions of the equation $\lambda\phi=A\phi$.

In the multidimensional case, in general, there does not exist a measure M such that the infinitesimal generator of a given continuous strong Markov process is symmetric in $L^2(M)$. However, in order to represent the infinitesimal generator, the measure M - occuring in (2) and (3) - plays an analogous role as m plays in the one-dimensional case. Accordingly, we prove the existence of transition densities with respect to such a measure M, see Theorem 4.1 and Theorem 5.1.

Let us mention that our result can be applied to processes with more general generators than the operators introduced by (3) or (in the one-dimensional case) the Feller-Dynkin operators (4) or their generalizations $\frac{dD_s^{+}\cdot - \cdot dk}{dm}$ (cf. K. Itô and H.P. McKean [10], 5.6). An example is given in section 6. This example shows that operators belonging to the class $\mathfrak{A}$ are not necessarily associated with harmonic functions forming a harmonic space. In particular, we allow the existence of absorbing points. Furthermore, considering the processes associated with our operator class $\mathfrak{A}$, we note that this class can be embedded in the theory of balayage spaces (cf. J. Bliedtner and W.Hansen [2], chapters II-IV). However, this is not our aim (it seems to be complicated to verify the very general axioms (B_1)-(B_4) (cf [2], II.4) or (H_1)-(H_5) (cf [2], III.1) in our concrete situation).

This paper surveys results proved by the author in the references [11] and [12]. However, Theorem 2.2 is closely related to classical results summarized e.g. by L. Stoica (cf. [14], chapter I).

2. LINEAR LOCAL OPERATORS AND ASSOCIATED CONTINUOUS STRONG MARKOV PROCESSES

We start with the definition of linear local operators in the sense of E.B. Dynkin, cf. [5], section V.4.

Let $\mathcal{D}\subseteq R^d$ be a domain. For any $x\in\mathcal{D}$ let $D_A(x)$ be a family of real functions defined on some neighbourhood of x such that to any pair (x,f), where $x\in\mathcal{D}$ and $f\in D_A(x)$, a real number $Af(x)$ can be attached. Suppose that $\tilde{f}=f$ on some neighbourhood of x and $f\in D_A(x)$ imply $\tilde{f}\in D_A(x)$ and $Af(x)=A\tilde{f}(x)$.

If $x\in\mathcal{D}$, $f_1,f_2\in D_A(x)$ and $c_1,c_2\in R$ then let $f=c_1f_1+c_2f_2$ belong to $D_A(x)$ and suppose, moreover, $Af(x)=c_1Af_1(x)+c_2Af_2(x)$.

Operators A satisfying these conditions we call *linear local operators*.

If $\mathcal{U}\subseteq\mathcal{D}$, $f\in D_A(x)$ for all $x\in\mathcal{U}$ and Af is continuous on $\mathcal{U}$ then we say that f belongs to $D_A(\mathcal{U})$.

It is well known that the Laplace operator Δ is a linear local operator. Moreover, it turns out that linear local operators share a lot of properties with the Laplacian. For this reason, we will use the notion of harmonic functions, harmonic measures, etc. also for the analogous objects defined for linear local operators:

We consider a certain class of linear local operators with the property that for any subdomain $\mathcal{U}$ of $\mathcal{D}$ the domain $D_A(\mathcal{U})$ is a linear subspace of $C(\mathcal{U})$ (the space of all continuous functions on $\mathcal{U}$).

For those operators introduce the following conditions:

(i) The *local maximum principle*: If for $x\in\mathcal{D}$ and $f\in D_A(x)$ the function f has a nonnegative local maximum at x then $Af(x)\le 0$.

Next we introduce the class of *regular* domains: Let $\mathcal{F}$ be a class of bounded domains $\mathcal{V}$ with $\bar{\mathcal{V}}\subset\mathcal{D}$ such that for any $x\in\mathcal{D}$ and any neighbourhood $\mathcal{N}$ of x there exists a domain $\mathcal{U}\in\mathcal{F}$ with $x\in\mathcal{U}\subseteq\mathcal{N}$. Moreover, suppose that for every compact set $K\subset\mathcal{D}$ a domain $\mathcal{U}\in\mathcal{F}$ with $K\subset\mathcal{U}$ can be found.

(ii) For every domain $\mathcal{U}\in\mathcal{F}$ there exists a function $g^{\mathcal{U}}\in D_A(\mathcal{U})$ with $Ag^{\mathcal{U}}(x)>0$, $x\in\mathcal{U}$.

Let $H_A(\mathcal{U})$ denote the space of all *harmonic functions* (with respect to A) on $\mathcal{U}\subseteq\mathcal{D}$, i.e.

$$H_A(\mathcal{U})=\{h\in D_A(\mathcal{U}): Ah(x)=0,\ x\in\mathcal{U}\}.$$

Let $C(\bar{\mathcal{U}})$ be the set of all continuous functions on $\mathcal{U}$ which have a continuous extension on $\bar{\mathcal{U}}$.

Define the spaces $H_A^{\beta}(\mathcal{U})$ and $\beta(\mathcal{U})=\beta_A(\mathcal{U})$ by

$$H_A^{\beta}(\mathcal{U})=H_A(\mathcal{U})\cap C(\bar{\mathcal{U}})$$

and

$$\beta(\mathcal{U})=\{\phi\in C(\partial\mathcal{U}):\phi=h|_{\partial\mathcal{U}},\ h\in H_A^{\beta}(\mathcal{U})\}.$$

Now we are able to introduce a third condition:

(iii) For any domain $\mathcal{U}\in\mathcal{F}$ the set $\beta(\mathcal{U})$ is dense in $C(\partial\mathcal{U})$.

Let a linear local operator belong to the *class* $\mathfrak{A}$ if it satisfies (i)-(iii). Throughout this section suppose $A\in\mathfrak{A}$.

Let $g^{\mathcal{U}}$, $\mathcal{U}\in\mathcal{F}$, be the functions mentioned in condition (ii) and let $\mu_{x,\mathcal{U}}$, $\mathcal{U}\in\mathcal{F}$, $x\in\mathcal{U}$, be the harmonic measures associated with $H_A(\mathcal{U})$. Suppose that the functions $g^{\mathcal{U}}$ are strictly subharmonic on $\mathcal{U}$, i.e. $g^{\mathcal{U}}(x)<\int_{\partial\mathcal{V}} g^{\mathcal{U}}(y)\ \mu_{x,\mathcal{V}}(dy)$ for any $\mathcal{V}\in\mathcal{F}$ with $\mathcal{V}\subseteq\mathcal{U}$ and $x\in\mathcal{V}$. (For example, this is satisfied if for all $\mathcal{U}\in\mathcal{F}$ the set $\beta(\mathcal{U})$ is the space of the continuous functions on $\partial\mathcal{U}$.)

For $\mathcal{U}\in\mathcal{F}$ and $x\in\mathcal{U}$ introduce the set $D_{\mathbb{A}}(x)$ of all real continuous functions f which are defined on some neighbourhood of x and have a limit

$$\mathbb{A}f(x)=\lim_{\mathcal{V}\downarrow x}\frac{\int_{\partial\mathcal{V}} f(y)\ \mu_{x,\mathcal{V}}(dy)-f(x)}{\frac{1}{Ag^{\mathcal{U}}(x)}\left(\int_{\partial\mathcal{V}} g^{\mathcal{U}}(y)\ \mu_{x,\mathcal{V}}(dy)-g^{\mathcal{U}}(x)\right)}.$$

(This limit is independent of $\mathcal{U}$.)

Linear local operators with the above properties have exactly the same representation as characteristic operators of continuous strong Markov processes:

Theorem 2.1. *Let* $\mathcal{U}\in\mathcal{F}$, $x\in\mathcal{U}$ *and* $f\in D_A(\mathcal{U})$. *Then* $f\in D_{\mathbb{A}}(x)$ *and*

$$Af(x)=\mathbb{A}f(x).$$

Furthermore, linear local operators belonging to the class $\mathfrak{A}$ induce continuous strong Markov processes, cf. P Courrège and P. Priouret [4] and L. Stoica [14], chapter I:

For $\mathcal{U}\in\mathcal{F}$ define

$$D_{\mathbf{A}\mathcal{U}}=\{f\in D_{\mathbb{A}}(\mathcal{U})\cap C_0(\mathcal{U}):\ \mathbb{A}f\in C_0(\mathcal{U})\}$$

and the operator $\mathbf{A}^{\mathcal{U}}:D_{\mathbf{A}^{\mathcal{U}}}\longrightarrow C_0(\mathcal{U})$ by

$$\mathbf{A}^{\mathcal{U}}f(x)=\mathbb{A}f(x),\ f\in D_{\mathbf{A}^{\mathcal{U}}},\ x\in\mathcal{U},$$

where $C_0(\mathcal{U})$ is the space of all continuous functions on $\mathcal{U}$ vanishing on $\partial\mathcal{U}$. Suppose that for all $\mathcal{U}\in\mathcal{F}$ the domain of definition $D_{\mathbf{A}^{\mathcal{U}}}$ is dense in $C_0(\mathcal{U})$ (a sufficient condition is given in [11], II.4 or [12], II.3).

Theorem 2.2 (a) *Let $\mathcal{U}\in\mathcal{F}$. The operator $\mathbf{A}^{\mathcal{U}}$ is the infinitesimal operator of a stochastically continuous $C_0(\mathcal{U})$-transition function $P^{\mathcal{U}}:[0,\infty)\times\mathcal{U}\times\mathcal{B}(\mathcal{U})\longrightarrow[0,1]$.*

(b) *There exists a continuous strong Markov process $X=(X_t,\zeta,\mathcal{M}_t,P_x)$ with state space $\mathcal{D}$ such that the $C_0(\mathcal{D})$-infinitesimal operator of the process $X^{\mathcal{U}}$ obtained by killing X on $\partial\mathcal{U}$ is $\mathbf{A}^{\mathcal{U}}$ ($\mathcal{U}\in\mathcal{F}$).*

Furthermore, the restriction of the characteristic operator of X to the continuous functions is $\mathbb{A}$.

3. LINEAR LOCAL OPERATORS WITH HARMONIC FUNCTIONS OF A SECOND ORDER ELLIPTIC DIFFERENTIAL OPERATOR

In this section we construct linear local operators which satisfy the local maximum principle (i) and have harmonic functions which coincide (in the context of section 2) with the harmonic functions of an elliptic operator

$$Bf(x)=b^{ij}(x)\frac{\partial^2}{\partial x^i \partial x^j}f(x)+b^i(x)\frac{\partial}{\partial x^i}f(x)+b(x)f(x).$$

(We regard the elliptic operator B as a linear local operator belonging to the class $\mathfrak{A}$.) Let us remark that this assumption is not so far from the general setting. Using probabilistic arguments A.V. Skorohod ([13], III.4, Theorem 8) has shown that - under weak conditions - the harmonic functions of the generator of a conservative continuous strong Markov process coincide with those of a second order elliptic differential operator, up to a (harmonic) space transformation. However, in section 6 we will give a simple example which cannot be treated in this way.

We consider two cases of elliptic operators B:

<u>Case I: Degenerated Elliptic Operators</u>

Let B be a degenerated elliptic operator of the form

$$Bf=\sum_{k=1}^{r} X_k^2 f+Yf+bf=b^{ij}\frac{\partial^2}{\partial x^i \partial x^j}f+b^i\frac{\partial}{\partial x^i}f+bf$$

where

$$X_k(x)=(\alpha_1^k(x),\dots,\alpha_d^k(x)),\ k=1,\dots,r,$$

and Y are vector fields of class $C^\infty(\mathcal{D})$ and $b\in C^\infty(\mathcal{D})$.

Suppose that the Lie algebra $\mathcal{L}(X_1,\dots,X_r)$ is for all $x\in\mathcal{D}$ of rank d and that

$$(b^{ij}(x))_{i,j=1,\dots,d}\neq\mathbf{0},\ x\in\mathcal{D},$$

(0 denotes the zero matrix). Finally, suppose the existence of $b_0<0$ such that

$$b(x)\leq b_0, \quad x\in\mathcal{D}.$$

<u>Case II: Non-degenerated Elliptic Operators</u>

Let B be an elliptic operator of the form

$$Bf=b^{ij}\frac{\partial^2}{\partial x^i\partial x^j}f+b^i\frac{\partial}{\partial x^i}f+bf$$

where $(b^{ij}(x))_{i,j=1,\ldots,d}$ is for all $x\in\mathcal{D}$ positive definite, $b^{ij}\in C^{(2,\lambda)}(\mathcal{D})$, $b^i\in C^{(1,\lambda)}(\mathcal{D})$ $(i,j=1,\ldots,d)$, $b\in C^{(0,\lambda)}(\mathcal{D})$ and $b(x)\leq 0$, $x\in\mathcal{D}$. (The superscript (k,λ) indicates Hölder continuity with exponent λ of the partial derivatives of order k.)

Subsequently, the cases I and II are treated simultaneously.

Let B^* denote the operator

$$B^*f=\frac{\partial^2}{\partial x^i\partial x^j}(b^{ij}f)-\frac{\partial}{\partial x^i}(b^if)+bf$$

and let $K^{\mathcal{U}}:\mathcal{U}\times\mathcal{U}\longrightarrow(0,\infty]$ be the Green's function of B on $\mathcal{U}\in\mathcal{F}$.

We remark that the operator B satisfies (i)-(iii).

Let A be a linear local operator belonging to the class $\mathfrak{A}$ satisfying the condition

(iii') For all $\mathcal{U}\in\mathcal{F}$ it holds

$$H_A(\mathcal{U})=H_B(\mathcal{U}).$$

Applying results of J.M. Bony [3] and R.-M. Hervé and M. Hervé [9] one can prove:

Theorem 3.1. *There exists a nonnegative locally finite measure* M *on* $\mathcal{D}$ *with*

$$\operatorname{supp} M=\bar{\mathcal{D}} \tag{5}$$

such that for all $\mathcal{U}\in\mathcal{F}$

(a) $f\in D_A(\mathcal{U})\cap C_0(\mathcal{U})$ *has a representation*

$$f(x)=-\int_{\mathcal{U}} K^{\mathcal{U}}(x,y)Af(y)\, M(dy),\ x\in\mathcal{U},$$

and

(b) $B^{\mathcal{U}}f(\phi):=\int_{\mathcal{U}} fB^{*}\phi\, dx=\int_{\mathcal{U}} \phi Af\, dM,\ \phi\in C_0^{\infty}(\mathcal{U}).$

($C_0^{\infty}(\mathcal{U})$ denotes the space of all $C^{\infty}(\mathcal{U})$-functions with compact support.)

Let us turn to a converse assertion of this theorem:

Introduce a nonnegative locally finite measure M on $\mathcal{D}$ satisfying (5) and

$$-\int_{\mathcal{U}} K^{\mathcal{U}}(\cdot,y)\, M(dy)\in C(\bar{\mathcal{U}}),\ \mathcal{U}\in\mathcal{F}. \tag{6}$$

If for $f\in C(\mathcal{U})$, $\mathcal{U}\in\mathcal{F}$, there exists $g=g(f)\in C(\bar{\mathcal{U}})$ such that for all $\phi\in C_0^{\infty}(\mathcal{U})$

$$\int_{\mathcal{U}} fB^{*}\phi\, dx=\int_{\mathcal{U}} \phi g\, dM$$

then let f belong to $D_{\Psi_{\mathcal{U}}}$. A linear local operator $\Psi=\Psi(M,B)$ can be introduced as follows:

$$D_{\Psi}(x)=\bigcup_{\{\mathcal{U}\in\mathcal{F}:x\in\mathcal{U}\}} D_{\Psi_{\mathcal{U}}},\ x\in\mathcal{D}, \tag{7}$$

and

$$\Psi f(x)=g(f)(x),\ f\in D_{\Psi_{\mathcal{U}}}, \tag{8}$$

for some $\mathcal{U}\in\mathcal{F}$ with $x\in\mathcal{U}$ (note that $f\in D_{\Psi}(x)$ implies the existence of $\mathcal{U}\in\mathcal{F}$ with $f\in D_{\Psi_{\mathcal{U}}}$).

Theorem 3.2. *The operator* Ψ *belongs to the class* $\mathfrak{A}$ *and satisfies* (iii').

Let us conclude this section with a general remark concerning the one-dimensional case:

If $B=\Delta$ then a theorem due to W. Feller [6] asserts that a linear local operator satisfying (i), (ii) and (iii') has a representation $D_m D_x^+ = \frac{d}{dm} D_x^+$. (Here D_x^+ denotes the right derivative and m is a locally finite measure.) Thus, the operator Ψ can be regarded as a multidimensional version of $D_m D_x^+$. In the reference [12], section IV.5, it is shown that $m=M$ implies $\Psi(M,\Delta)=D_m D_x^+$.

4. TRANSITION DENSITIES OF CONTINUOUS STRONG MARKOV PROCESSES

Let B be an operator given in section 3, let M be a measure satisfying (5) and (6) and let Ψ be the operator defined by (7) and (8). Furthermore, let X be the associated process (introduced in Theorem 2.2) and let P denote its transition function.

Theorem 4.1. *For all* $x\in\mathcal{D}$ *and all* $t>0$ *there exists* *a function* p(t,x,.) *such that for any Borel set* $\Gamma\subseteq\mathcal{U}$

$$P(t,x,\Gamma)=\int_\Gamma p(t,x,y)\ M(dy).$$

Since the operator Ψ is identified as a multidimensional version of $D_m D_x^+$ (section 3) Theorem 4.1 extends a well known result for one-dimensional continuous strong Markov processes (see K. Itô and H.P. McKean [10], 4.11) to the multidimensional case.

Finally, we mention that for symmetric processes a similar result is given by M. Fukushima [7], 4.2 and 4.3.

5. A GENERALIZATION

Let us consider a more general situation than in section 3. Suppose (i)-(iii) and suppose that for any domain $\mathcal{U}\in\mathcal{F}$ and any function $g\in C(\bar{\mathcal{U}})$ there exists a function $f\in D_A(\mathcal{U})\cap C_0(\mathcal{U})$ with $Af=g$.

Let $H_A^e(\mathcal{U})$ denote the completion of H_A^β with respect to the uniform convergence. We introduce the following condition:

(iv) For all $\mathcal{U}\in\mathcal{F}$ there exists a function $f^o=f^o(\mathcal{U})\in D_A(\mathcal{U})\cap C_0(\mathcal{U})$ with $Af^o(x)<0$, $x\in\bar{\mathcal{U}}$, a nonnegative measurable function $\mathbb{K}^{\mathcal{U}}$ on $\mathcal{U}\times\mathcal{U}$ and a signed measure $M^{\mathcal{U}}_{f^o}$ of finite variation such that:

1) $$f^o(x)=-\int_{\mathcal{U}} \mathbb{K}^{\mathcal{U}}(x,y)\ M^{\mathcal{U}}_{f^o}(dy),\ x\in\mathcal{U}. \qquad (9)$$

2) For arbitrary domains $\mathcal{M},\mathcal{N}\subseteq\mathcal{U}$ with $\bar{\mathcal{N}}\subset\mathcal{M}$ the

function $\mathbb{K}^{\mathcal{U}}$ is bounded on $\mathcal{N}\times(\mathcal{U}\setminus\mathcal{M})$ and for any $y\in\mathcal{U}$ the function $\mathbb{K}^{\mathcal{U}}(.,y)$ is lower semi-continuous on $\mathcal{U}$.

3) For all $y\in\mathcal{U}$ and all $\mathcal{V}\in\mathcal{F}$ with $\bar{\mathcal{V}}\subset\mathcal{U}\setminus\{y\}$ the restriction of $\mathbb{K}^{\mathcal{U}}(.,y)$ to the set $\mathcal{V}$ belongs to $H_A^e(\mathcal{U})$.

This is a rather technical condition but it leads us to the following assertion:

Theorem 5.1. (a) *For all* $\mathcal{U}\in\mathcal{F}$, *there exists a finite measure* $M^{\mathcal{U}}$ *on* $\mathcal{U}$ *with* supp $M^{\mathcal{U}}=\bar{\mathcal{U}}$ *and* $-\int_{\mathcal{U}} \mathbb{K}^{\mathcal{U}}(.,y)\ M^{\mathcal{U}}(dy)\in C(\bar{\mathcal{U}})$ *such that any function* $f\in D_A(\mathcal{U})\cap C_0(\mathcal{U})$ *has a representation*

$$f(x)=-\int_{\mathcal{U}} \mathbb{K}^{\mathcal{U}}(x,y)Af(y)\ M^{\mathcal{U}}(dy),\ x\in\mathcal{U}.$$

(b) *Suppose the existence of a function* $\mathbb{K}:\mathcal{D}\times\mathcal{D}\longrightarrow(0,\infty]$ *such that for* $\mathcal{U}\in\mathcal{F}$ *and* $y\in\mathcal{U}$

$$\mathbb{K}(.,y)|_{\mathcal{U}}-\mathbb{K}^{\mathcal{U}}(.,y)$$

belongs to $H_A^e(\mathcal{U})$. *Then there exists a measure* M *on* $\mathcal{D}$ *which coincides (on all domains* $\mathcal{U}\in\mathcal{F}$) *with the measure* $M^{\mathcal{U}}$.

Furthermore, the associated Markov process admits transition densities with respect to the measure M *(in the sense of* Theorem 4.1).

It turns out that, under weak additional conditions, a linear local operator satisfies condition (iv) whenever it is associated with a continuous strong Markov process admitting transition densities (cf. [12], III.3).

Let us remark that, in order to prove Theorem 3.1, and Theorem 4.1 (via Theorem 5.1), we have introduced the condition (iv) (see reference [12], chapter IV). However, applying Theorem 5.1 (i.e. verifying condition (iv)), one can investigate classes of operators and associated processes which cannot be treated in the context of section 3. An example is given in section 6.

Representations of the form (9) have been considered by R.-M. Hervé in the fundamental paper [8], section 17 (Theorem 17.1). The system of axioms introduced there is sufficient to verify our condition (iv). However, this system is not completely compatible with our conditions (i)-(iii). For example, in the above reference for all $\mathcal{U}\in\mathcal{F}$ and all nonnegative $h\in H_A(\mathcal{U})$ the alternative $h>0$ on $\mathcal{U}$ or $h\equiv 0$ is supposed.

Furthermore, in order to derive Theorem 5.1, we have not assumed that (for all points $y\in\mathcal{U}$) all potentials, the support of which is y, are proportional (cas de l' unicité, cf. R.-M. Hervé [8], section 18). In the following section these differences to the classical theory will be discussed.

6. A NON-STANDARD EXAMPLE

We conclude this note with an example of an operator $A\in\mathfrak{A}$ satisfying condition (iv) the harmonic functions of which do not form a harmonic space in the sense of H. Bauer and J. Bliedtner and W. Hansen (cf. [1], section I.1, and [2], sections III.7 and VIII.1). In particular, we show the existence of a point $x\in\mathcal{D}$ with $h(x)=0$ for all $\mathcal{U}\in\mathcal{F}$ with $x\in\mathcal{U}$ and all h belonging to $H_A^e(\mathcal{U})$; i.e. x is an absorbing point.

Moreover, we represent this example in the form (2) and show that the associated continuous strong Markov process admits transition densities with respect to a canonical measure.

Let $\mathcal{B}_r\subset R^2$ denote the ball with centre (0,0) and radius r. Furthermore, set $\mathcal{I}=\{(x_1,0):x_1\in(1,2)\}$ and put $\mathcal{P}=(\frac{3}{2},0)$. Let $\mathcal{D}=\mathcal{B}_2\setminus\overline{\mathcal{B}_1}$ and let $D_A(x)$, $x\in\mathcal{D}$, be the set of all real functions defined on some neighbourhood of x such that the number

$$Af(x)=\begin{cases} \frac{1}{4}\Delta f(x) & x\in\mathcal{D}\setminus\mathcal{I} \\ \frac{1}{2}\frac{\partial^2}{\partial x_1^2}f(x) & x\in\mathcal{I}\setminus\mathcal{P} \\ -f(x) & x=\mathcal{P} \end{cases}$$

(where $\Delta=\frac{\partial^2}{\partial x_1^2}+\frac{\partial^2}{\partial x_2^2}$) exists. By the procedure given in section 2 introduce a linear local operator A:

Let $\mathcal{F}$ be the set of all domains $\mathcal{U}$ with smooth boundary such that $\bar{\mathcal{U}}\subset\mathcal{D}$. Obviously, the operator A satisfies the local maximum principle (i).

Furthermore, for $g(x)=x_1^2+x_2^2-\frac{5}{4}$, $x\in\mathcal{D}$, it holds $Ag(x)=1$, $x\in\mathcal{D}$. Thus, we have (ii).

Let $\mathcal{U}\in\mathcal{F}$ and $\phi\in C(\partial\mathcal{U})$. If $\mathcal{U}\cap\mathcal{I}\neq\emptyset$ then $\mathcal{U}\cap(\mathcal{I}\setminus\mathcal{P})$ consists of a countable number of sets $\mathcal{I}_n=\{(x_1,0):a_n<x_1<b_n)\}$ which we call the components of $\mathcal{U}\cap(\mathcal{I}\setminus\mathcal{P})$. Consider the solution $\tilde{h}$ of the classical Dirichlet problem on $\mathcal{U}\setminus\mathcal{I}$ with the

continuous boundary function $\tilde{\phi}$ on $\partial(\mathcal{U}\setminus\mathcal{I})$, given by $\tilde{\phi}=\phi$ on $\partial\mathcal{U}$, $\tilde{\phi}(\mathcal{P})=0$ (if $\mathcal{P}\in\mathcal{U}$) and $\tilde{\phi}$ linear on the components of $\mathcal{U}\cap(\mathcal{I}\setminus\mathcal{P})$ (if $\mathcal{U}\cap\mathcal{I}\neq\emptyset$). Now, the fuction

$$h=\begin{cases} \tilde{h} \text{ on } \mathcal{U}\setminus\mathcal{I} \\ \tilde{\phi} \text{ on } \mathcal{U}\cap\mathcal{I} \end{cases}$$

satisfies $Ah=0$ on $\mathcal{U}$ and $h|_{\partial\mathcal{U}}=\phi$. Consequently, we can state the validity of condition (iii). Thus, $A\in\mathfrak{A}$.

For all $\mathcal{U}\in\mathcal{F}$ with $\mathcal{P}\in\mathcal{U}$ and all $h\in H_A^{\beta}(\mathcal{U})(=H_A(\mathcal{U}))$ we have $h(\mathcal{P})=0$, i.e. $\mathcal{P}$ is an absorbing point. Consequently, the harmonic functions of A do not correspond to some harmonic space in the sense of the above references. Moreover, for a nonnegative $h\in H_A(\mathcal{U})$ the alternative $h>0$ on $\mathcal{U}$ or $h\equiv 0$ does not hold (recall the comments given in section 5).

Let $\tilde{p}_{\alpha,\beta}$ be the solution of the classical Dirichlet problem on $\mathcal{D}\setminus\mathcal{I}$ with boundary function ϕ on $\partial(\mathcal{D}\setminus\mathcal{I})$, given by $\phi=0$ on $\partial\mathcal{D}$ and $\phi((y_1,0))=\alpha(y_1-1)$, $1<y_1\leq\frac{3}{2}$, and $\phi((y_1,0))=\beta(2-y_1)$, $\frac{3}{2}<y_1<2$, on $\mathcal{I}$, $0<\alpha\leq\beta$. The functions

$$p_{\alpha,\beta}=\begin{cases} \tilde{p}_{\alpha,\beta} & \text{on } \mathcal{D} \\ \phi & \text{on } \mathcal{I} \end{cases}$$

are potentials (in the sense of R.-M. Hervé [8]) the support of which is the point $\mathcal{P}$. Therefore, there exist potentials, the support of which is the point $\mathcal{P}$, which are not proportional.

According to Theorem 2.1 for $x\in\mathcal{D}$ and $f\in D_A(x)$ we have a representation (1).

Now, we are interested in a potential

representation (2). To this end, we apply Theorem 5.1. Thus, we have to search for functions $f^o=f^o(\mathcal{U})$ and $\mathbb{K}^{\mathcal{U}}$ and measures $M^{\mathcal{U}}$, $\mathcal{U}\in\mathcal{F}$, such that condition (iv) is satisfied.

Let $G^{\mathcal{D}\setminus\mathcal{I}}$ denote the classical Green's function on $\mathcal{D}\setminus\mathcal{I}$ and let $G^{(1,\frac{3}{2})}$ and $G^{(\frac{3}{2},2)}$ denote the classical Green's functions on $(1,\frac{3}{2})$ and $(\frac{3}{2},2)$. Furthermore, for $y\in(1,\frac{3}{2})$ let K^y denote that function belonging to $C_0(\mathcal{D})$ which satisfies $\Delta K^y=0$ on $\mathcal{D}\setminus\mathcal{I}$ and $K^y(x)=0$, $x\in[\frac{3}{2},2)$ and coincides on $(1,\frac{3}{2})$ with $G^{(1,\frac{3}{2})}(\cdot,y)$. Moreover, for $y\in(\frac{3}{2},2)$ let K^y denote that function belonging to $C_0(\mathcal{D})$ which satisfies $\Delta K^y=0$ on $\mathcal{D}\setminus\mathcal{I}$ and $K^y(x)=0$, $x\in(1,\frac{3}{2}]$ and coincides on $(\frac{3}{2},2)$ with $G^{(\frac{3}{2},2)}(\cdot,y)$.

Next define

$$\mathbb{K}(\cdot,y)=\begin{cases}\frac{1}{4}G^{\mathcal{D}\setminus\mathcal{I}}(\cdot,y) & y\in\mathcal{D}\setminus\mathcal{I}\\ \frac{1}{2}K^y & y\in\mathcal{I}\setminus\mathcal{P}\\ p_{2,2} & y=\mathcal{P}\end{cases}$$

(for $x\in\mathcal{I}$ set $G^{\mathcal{D}\setminus\mathcal{I}}(x,y)=0$) and for any Borel set $\Gamma\subseteq\mathcal{D}$ put

$$M(\Gamma)=\Lambda^2(\Gamma)+\Lambda^1(\Gamma\cap\mathcal{I})+\chi_\Gamma(0) \tag{10}$$

where Λ^d denotes the Lebesgue measure in R^d and χ is the indicator function.

Let $\mathcal{U}\in\mathcal{F}$. For $y\in\mathcal{U}$ let $h^y_{\mathcal{U}}\in H_A(\mathcal{U})$ be the (uniquely determined) function satisfying $h^y_{\mathcal{U}}(z)=\mathbb{K}(z,y)$, $z\in\partial\mathcal{U}$. Setting

$$\mathbb{K}^{\mathcal{U}}(x,y)=\mathbb{K}(x,y)-h^y_{\mathcal{U}}(x),\ x,y\in\mathcal{U},$$

and

$$f^{o}(x)=\int_{\mathcal{U}} \mathbb{K}^{\mathcal{U}}(x,y)\ M(dy).$$

one verifies the validity of condition (iv). Now, by Theorem 3.1(a), there exists a representation in the form (2). Indeed, for $f \in D_A(\mathcal{U}) \cap C_0(\mathcal{U})$ we have the representation

$$f(x)=-\int_{\mathcal{U}} \mathbb{K}^{\mathcal{U}}(x,y)Af(y)\ M(dy),\ x \in \mathcal{U},$$

where **M** is the measure introduced by (10).

Let us mention that there does not exist a representation in the form (3) with coefficients introduced in section 3. (In these cases absorbing points do not exist.)

The process $X=(X_t,\zeta,\mathcal{M}_t,P_x)$ associated with the operator A (by Theorem 2.2) moves as a two-dimensional Brownian motion until it reaches $\mathcal{I}$ or is killed on $\partial\mathcal{D}$. If the process reaches $\mathcal{I}$ then it moves there as a one-dimensional Brownian motion until it hits the point $\mathcal{P}$ or is killed on $\partial\mathcal{I}=\{(1,0),(2,0)\}$. If the process hits $\mathcal{P}$ then it waits there an exponentially distributed period (with parameter 1) and is killed after this period.

Finally, Theorem 5.1 asserts that *the transition probabilities of X admit densities with respect to the measure M.*

REFERENCES

1. H. Bauer, *Harmonische Räume und ihre Potentialtheorie*, Lecture Notes in Mathematics **22**, Springer, Berlin Heidelberg New York, 1966.
2. J. Bliedtner, W. Hansen, *Potential theory, An analytic and probabilistic approach to balayage*, Springer, Berlin Heidelberg New York Tokyo, 1986.
3. J.M. Bony, Principe du maximum et inégalité de Harnack pour les opérateurs elliptiques dégénérés, in *Séminaire de la théorie du potentiel*. IHP, **12**e année, no 10 (1967-1968).
4. P. Courrège, P. Priouret, Recollements de processus de Markov. *Publ. Inst. Stat. Univ. Paris*, **14** (1965), 275-377.
5. E.B. Dynkin, *Markov processes* I,II. Springer, Berlin Göttingen Heidelberg 1965.
6. W.Feller, Differential operators with the positive maximum property. *Ill. J. Math.* **3** (1959), 182-185.
7. M. Fukushima, *Dirichlet forms and Markov processes*. North Holland/Kodansha, Tokyo, 1980.
8. R.-M. Hervé, Recherches axiomatiques sur la théorie des functions surharmoniques et du potentiel, Ann. Inst. Fourier **12** (1962), 415-571.
9. R.-M. Hervé, M. Hervé, Les functions surharmoniques dans l'axiomatique de M. Brelot associées a un opérateur elliptique dégénéré. *Ann. Inst. Fourier* **22**, 2 (1972), 131-145.
10. K. Itô, H.P. McKean, *Diffusion processes and their sample path*. 2nd edn, Springer, Berlin Heidelberg New York, 1974.
11. J.-U. Löbus, Linear local operators satisfying the local maximum principle and associated Markov processes and geometrical structures in R^d I-IV, *Preprint FSU Jena*, Math/91/4, 1991.
12. J.-U. Löbus, *Generalized diffusion operators.* to appear: Mathematical Research, Akademie

Verlag Berlin, 1993.

13. I.I. Gihman, A.V. Skorohod, *The theory of stochastic processes*, III, Springer, Berlin Heidelberg New York, 1979.

14. L. Stoica, *Local operators and Markov processes*. Lecture Notes in Mathematics **816**, Springer, Berlin Heidelberg New York, 1980.

EXISTENCE OF SOLUTIONS OF STOCHASTIC REACTION–DIFFUSION EQUATIONS WITH CONTINUOUS REACTION

RALF MANTHEY and KATRIN MITTMANN

Friedrich–Schiller–Universität, Mathematische Fakultät, UHH, 17. OG, D–O–6900 Jena, Germany

Abstract We study stochastic parabolic differential equations with a Gaussian noise which is white in time and with a space-time white noise respectively. For such equations we give definitions of solutions and show the existence of this solutions under quite general conditions.

0. INTRODUCTION

Consider the parabolic differential problem

$$\begin{aligned} \mathbf{L}u(t,x) &= \alpha(t,x,u(t,x)), & (t,x) \in (0,T] \times Q, \\ u(0,x) &= \vartheta(x), & x \in \overline{Q}, \end{aligned}$$

with Neumann boundary conditions

$$\tfrac{\partial}{\partial \nu} u(t,x) = 0, \qquad (t,x) \in (0,T] \times \partial Q,$$

or Dirichlet boundary conditions

$$u(t,x) = 0, \qquad (t,x) \in (0,T] \times \partial Q,$$

where $\mathbf{L} = \frac{\partial}{\partial t} - A$ with

$$A(t,x) = \sum_{i,k=1}^{d} a_{ik}(t,x) \frac{\partial^2}{\partial x_i \partial x_k} + \sum_{i=1}^{d} b_i(t,x) \frac{\partial}{\partial x_i} + c(t,x)$$

is a strongly parabolic second order differential operator with Hölder continuous coefficients [1]. We assume that $T > 0$ is an arbitrary but fixed real number. By Q we denote an open convex bounded set in $\mathbb{R}^d$ with a sufficiently smooth boundary ∂Q. More precisely, we assume that ∂Q satisfies the so-called Ljapunov condition [11]. For simplicity we require that Q is a subset of $[0,a]^d := \{x \in \mathbb{R}_+^d : 0 \le x_i \le a, i = 1, ..., d\}$ for a certain $a > 0$; obviously, this is not a restriction. The initial function $\vartheta : \overline{Q} \times \Omega \to \mathbb{R}$ is assumed to be continuous and satisfies $\vartheta/_{\partial Q} \equiv 0$ in the case of Dirichlet boundary conditions, and $\alpha : [0,T] \times \overline{Q} \times \mathbb{R} \to \mathbb{R}$ is supposed to be Borel measurable. Finally, the symbol $\frac{\partial}{\partial \nu}$ stands for the derivative with respect to the inner co-normal.

It is well-known that parabolic equations of the considered type can be succesfully used to model real processes like the time evolution of a spatial distributed population or a density of a chemical substance in a reaction-diffusion system and many others (D. Henry [4]). For our further investigations let $(\Omega, \mathcal{F}, \mathbb{P})$ be a complete probability space. In the above mentioned real processes one usually observes the existence of random disturbances. We shall model this random influence by an additive noise over $[0,T] \times [0,a]^d$ with an intensity $\beta : [0,T] \times \overline{Q} \times \Omega \times \mathbb{R} \to \mathbb{R}$. For the physical background see the papers of P. C. Hohenberg and B. I. Halperin [5] or S. Ma [6]. By the central limit theorem argument [2] we can reduce our consideration to the Gaussian case. If we assume that the space-time correlations of the (microscopic) noise are much smaller than those of the (macroscopic) solution process u, we can ignore them by taking a white noise. If the size of the space correlations is comparable, we choose a noise which is coloured in space.

By this considerations we arrive at the reaction-diffusion equation

$$\begin{aligned} \mathbf{L}u(t,x) &= \alpha(t,x,u(t,x)) + \beta(t,x,u(t,x)) \cdot \nu(t,x), \\ &\qquad (t,x) \in (0,T] \times Q, \end{aligned}$$

with the corresponding boundary conditions and a random initial condition. Here, ν stands for a Gaussian noise considered above. In the first section we study a formal Dirichlet problem with a Gaussian noise which is white in time and coloured in space. In section 2 we consider a formal Neumann problem with a Gaussian space-time white noise. Finally, in the last section we deal with the Cauchy problem corresponding to the problem considered in section 2.

1. MULTIDIMENSIONAL STOCHASTIC REACTION–DIFFUSION EQUATIONS WITH DIRICHLET BOUNDARY CONDITIONS

In this section we consider the Dirichlet problem (**D**)

$$\begin{aligned} \mathrm{L}u(t,x) &= \alpha(t,x,u(t,x)) + \beta(t,x)\xi(t,x), \\ & \qquad\qquad\qquad\qquad (t,x) \in (0,T] \times Q, \\ u(t,x) &= 0, \qquad\qquad\qquad (t,x) \in [0,T] \times \partial Q, \\ u(0,x) &= \vartheta(x), \qquad\qquad\qquad x \in \overline{Q}. \end{aligned}$$

Here the mapping $\beta : [0,T] \times \overline{Q} \times \Omega \to \mathbb{R}$ is a measurable and bounded random function. We assume that the initial function $\vartheta : \overline{Q} \times \Omega \to \mathbb{R}$ is pathwise continuous and satisfies $\vartheta/_{\partial Q} \equiv 0$. Finally, ξ denotes a Gaussian noise which is white in time and coloured in space. That means, $\xi = (\xi(\phi))_{\phi \in \mathcal{D}((0,T)\times(0,a)^d)}$ is a Gaussian generalized random field in the sense of I. M. Gelfand and N. J. Wilenkin [3] with

$$\mathbb{E}\xi(\phi) = 0$$

and

$$\mathbb{E}\xi(\phi)\xi(\psi) = \int_0^T \int_{[0,a]^d} \int_{[0,a]^d} \phi(t,x)\psi(t,y)\mu(x,y)\,dxdydt$$

for any $\phi, \psi \in \mathcal{D}((0,T) \times (0,a)^d)$, where $\mathcal{D}$ denotes the test function space. In this paper the mapping μ is given by

$$\mu(x,y) := \sum_{k=1}^{\infty} \lambda_k h_k(x) h_k(y), \qquad x,y \in [0,a]^d,$$

where $(\lambda_k)_{k\in\mathbb{N}}$ is a summable sequence of nonnegative real numbers, and $(h_k)_{k\in\mathbb{N}}$ denotes an orthonormal basis of trigonometric functions in $L_2((0,a)^d)$.

In order to formulate what we mean by a solution of the formal Dirichlet problem under consideration, we need a stochastic integral we shall shortly introduce now. It is easy to show that there exists a Wiener process B on $L_2((0,a)^d)$ with covariance operator Γ defined by

$$\Gamma h := \sum_{k=1}^{\infty} \lambda_k < h, h_k > h_k, \qquad h \in L_2((0,a)^d),$$

where $< .,. >$ denotes the inner product in $L_2((0,a)^d)$. We always asssume that ϑ and the family $(B_t(h_k))_{t\in[0,T],k\in\mathbb{N}}$ are independent. We define a so-called standard filtration $\mathbb{F} = (\mathcal{F}_t)_{t\geq 0}$ by setting

$$\mathcal{F}_t := \sigma\{\vartheta(x), B_s(h_k) : x \in \overline{Q},\ s \leq t,\ k \in \mathbb{N}\}_0^+,$$

i.e., $\mathbb{F}$ is right continuous and complete. Let $\mathcal{B}(X)$ denote the Borel σ-algebra on a metric space X. For $f : [0,T] \times [0,a]^d \times \Omega \to \mathbb{R}$ with the property that $f/_{[0,s]\times[0,a]^d\times\Omega}$ is $\mathcal{B}([0,s])\otimes\mathcal{B}([0,a]^d)\otimes\mathcal{F}_s$-measurable for any $s \in [0,T]$ and such that

$$\mathbb{E}\int_0^T \left[\int_{[0,a]^d} |f(s,x)|\,dx\right]^2 dt \ < \ \infty$$

holds, we define the stochastic integral

$$\int_{[0,t]\times[0,a]^d} f(s,x)\,dB_s(x)dx \ := \ \sum_{k=1}^{\infty}\int_0^t\int_{[0,a]^d} f(s,x)h_k(x)\,dx\,dB_s(h_k).$$

It is easy to check that the moment condition introduced above ensures the mean-square convergence of the right-hand side. If f possesses the measurability condition required to define this stochastic integral we shall say that f is *progressively measurable.*

One can show, that there exists a phathwise continuous Gaussian random field $W = (W(t,x))_{(t,x)\in[0,T]\times[0,a]^d}$ with the following properties.

The derivative (in the sense of generalized random fields) $\frac{\partial}{\partial t}\frac{\partial}{\partial x_1}\cdots\frac{\partial}{\partial x_d}\Phi_W$ of the Gaussian generalized random field Φ_W which is given by

$$\Phi_W(\phi) \ := \ \int_0^T\int_{[0,a]^d} W(t,x)\phi(t,x)\,dxdt, \quad \phi \in \mathcal{D}((0,T)\times(0,a)^d),$$

coincides in distribution with ξ.

For all functions $f : [0,T] \times [0,a]^d \times \Omega \to \mathbb{R}$ with the above mentioned properties one can define the stochastic integrals

$$\int_{[0,t]\times[0,a]^d} f(s,x)\,dW(s,x),$$

$t \in [0,T]$, such that these integrals and the integrals with respect to B coincide $\mathbb{P}$-a.s.

In the sequel β is always assumed to be progressively measurable.

1.1 DEFINITION (*strongly measurable solution*). *A random field $u : [0,T] \times \overline{Q} \times \Omega \to \mathbb{R}$ is called a strongly measurable solution to* **(D)** *if the following holds:*

(i) $u(t,.)$ *is $\mathcal{F}_t$-measurable for any $t \in [0,T]$.*

(ii) u *is pathwise continuous.*

(iii) u *satisfies*

$$u(0,x) = \vartheta(x)$$

and

$$\begin{aligned} u(t,x) &= \int_Q G(t,x;0,y)\vartheta(y)\,dy \\ &+ \int_0^t \int_Q G(t,x;s,y)\alpha(s,y,u(s,y))\,dyds \\ &+ \int_0^t \int_{[0,a]^d} G(t,x;s,y)\beta(s,y)\mathbf{1}_Q(y)\,dB_s(y)dy \end{aligned}$$

$\mathbb{P}$*-a.s. for all $(t,x) \in [0,T] \times \overline{Q}$.*

Here, by G we denote the Green's function corresponding to **L** on $[0,T] \times \overline{Q}$ with Dirichlet boundary conditions [11].

1.2 DEFINITION (*weakly measurable solution*). *A random field $u : [0, T \times \overline{Q} \times \Omega \to \mathbb{R}$ is called a weakly measurable solution to* **(D)** *if* **(ii)**, **(iii)** *of the previous definition and the following hold.*

(i*) $u(t,.)$ *is $\mathcal{F}_T$-measurable for any $t \in [0,T]$.*

Now we formulate some results on existence, uniqueness and comparison of solutions to **(D)**.

1.3 THEOREM. *Let α Lipschitz continuous in u, i.e., there exists a constant $L > 0$ such that*

$$|\alpha(t,x,u) - \alpha(t,x,v)| \leq L \cdot |u - v|,$$

holds for all $(t,x) \in [0,T] \times \overline{Q}, u, v \in \mathbb{R}$, and bounded. Assume

$$\sup_{x \in \overline{Q}} \mathbb{E}|\vartheta(x)|^2 < \infty.$$

Then there exists a pathwise unique strongly measurable solution to **(D)**.

To prove theorem 1.3 one can use standard arguments which are explained in the papers of R. Manthey [7,8].

In the following we only consider differential operators $\mathbf{L}$ with the property $c(t,x) \leq 0$ for all $(t,x) \in [0,T] \times \overline{Q}$. This is not a restriction because **(D)** can always be reduced to a problem such that this condition is fulfilled. For this we could choose

$$\tilde{\mathbf{L}}(t,x) \;=\; \mathbf{L}(t,x) + c(t,x)$$

and

$$\tilde{\alpha}(t,x,u) \;=\; \alpha(t,x,u) - c(t,x).$$

1.4 THEOREM. *Let α_1 and α_2 be Lipschitz continuous in $u \in \mathbb{R}$, Hölder continuous in $(t,x) \in [0,T] \times \overline{Q}$ uniformly in $u \in \mathbb{R}$, and suppose that*

$$\alpha_1(t,x,u) \;\geq\; \alpha_2(t,x,u)$$

holds for all $(t,x,u) \in [0,T] \times \overline{Q} \times \mathbb{R}$.
Assume that u_i, $i = 1,2$, are solutions to **(D)** *corresponding to (α_i, β), $i = 1,2$, with the same initial condition. Then*

$$u_1(t,x) \;\geq\; u_2(t,x)$$

holds for all $(t,x) \in [0,T] \times \overline{Q}$ $\mathbb{P}$-a.s.

Idea of the proof.
We define v by setting $v := u_1 - u_2$ and α by

$$\alpha(t,x) \;:=\; \alpha_1(t,x,u_1(t,x)) - \alpha_2(t,x,u_2(t,x)).$$

It is easy to show that v solves the Dirichlet problem

$$\begin{aligned}
\mathbf{L}v(t,x) &= \frac{\partial}{\partial t}v(t,x) - A(t,x)v(t,x) = \alpha(t,x), && (t,x) \in (0,T] \times \overline{Q}, \\
v(0,x) &= 0, && x \in \overline{Q}, \\
v(t,x) &= 0, && t \in [0,T], x \in \partial Q.
\end{aligned}$$

for $\mathbb{P}$-a.a. $\omega \in \Omega$. By the help of the deterministic theory one can show that

$$v(.,.,\omega) \geq 0$$

holds for $\mathbb{P}$-a.a. $\omega \in \Omega$. This ends the proof of theorem 1.4.

In order to show the next theorem we need the following well-known appproximation lemma.

1.5 LEMMA. *Let Y be a normed space and X be a closed subset of Y. For any continuous and bounded function $f : X \to \mathbb{R}$ there exist sequences $(a_n)_{n\geq 1}$ and $(b_n)_{n\geq 1}$, $a_n, b_n : X \to \mathbb{R}$, $n = 1, 2, \ldots$, of Lipschitz continuous mappings such that $a_n \uparrow f$ and $b_n \downarrow f$ pointwise. The functions $|a_n|$, $n \in \mathbb{N}$, and $|b_n|$, $n \in \mathbb{N}$, are bounded by the same constant as $|f|$.*

Examples of such functions are given by $b_n(x) = \sup_{z\in X}\{f(z) - n\cdot \mid x - z \mid\}$ and $a_n(x) = \inf_{z\in X}\{f(z) - n\cdot \mid x - z \mid\}$, where $\mid . \mid$ denotes the norm in Y.

1.6 THEOREM. *Suppose that the initial function satisfies*

$$\sup_{x\in\overline{Q}} \mathbb{E}|\vartheta(x)|^2 < \infty.$$

(i) *Let $\alpha : [0,T] \times \overline{Q} \times \mathbb{R} \to \mathbb{R}$ be continuous. Assume that there exist constants $c^+ \geq 0$ and $c^- \leq 0$ such that*

$$\alpha(t,x,u) \geq c^-(1-u)$$

holds for all $u \leq 0$ and all (t,x) and

$$\alpha(t,x,u) \leq c^+(1+u)$$

holds for all $u \geq 0$ and all (t,x).
Then there exists at least one weakly measurable solution to **(D)**.

(ii) *If α is continuous and bounded then there exists at least one strongly measurable solution to* **(D)**.

(iii) *If α satisfies the condition (i) and if it is local Lipschitz continuous in u, i.e., for every $N > 0$ there exists a constant $L(N) > 0$ such that*

$$|\alpha(t,x,u) - \alpha(t,x,v)| \leq L(N) \cdot |u - v|,$$

holds for all $(t,x) \in [0,T]\times\overline{Q}$ and all $u, v \in \mathbb{R}$ with $|u|, |v| \leq N$, then there exists a pathwise unique strongly measurable solution to **(D)**.

Idea of the proof.
(ii) If α is bounded, then by lemma 1.5 there exist Lipschitz continuous functions α_N with $\alpha_N \uparrow \alpha$, $N \to \infty$. Let u_N be the solutions to **(D)** corresponding to $(\vartheta, \alpha_N, \beta)$, $N \in \mathbb{N}$. From theorem 1.4 we get $u_N \leq u_{N+1}$, $N \in \mathbb{N}$, $\mathbb{P}$-a.s. Furthermore, one can show, that

$$\sup_{t,x,N} |u_N(t,x)| \leq \eta$$

with a $\mathbb{P}$-a.s. bounded random variable η holds. Therefore there exists the pointwise limit

$$u(t,x,\omega) = \lim_{N\to\infty} u_N(t,x,\omega)$$

on $[0,T] \times \overline{Q} \times \tilde{\Omega}$, where $\tilde{\Omega} \subset \Omega$ has $\mathbb{P}$-measure one. It is not difficult to show that u is a strongly measurable solution to **(D)**.
(i) For $M \in \mathbb{N}$ we define

$$\alpha_M(t,x,v) := \begin{cases} \alpha(t,x,v) & ; \ |v| \leq M \\ \alpha(t,x,M\,\mathrm{sgn}(v)) & ; \ |v| > M \end{cases}$$

with

$$\mathrm{sgn}(v) := \begin{cases} 1 & ; \ v \geq 0 \\ -1 & ; \ v < 0. \end{cases}$$

By **(ii)** there exist strongly measurable solutions u_M to **(D)** corresponding to $(\vartheta, \alpha_M, \beta)$, $M \in \mathbb{N}$. One can show that for $\mathbb{P}$-a.a. $\omega \in \Omega$ there exists a constant $K(\omega)$ such that

$$\sup_{t,x,M} |u_M(t,x,\omega)| \leq K(\omega)$$

holds. For these ω we set

$$u(t,x,\omega) = u_M(t,x,\omega),$$

if

$$M-1 \leq \sup_{s,y,N} |u_N(s,y,\omega)| < M$$

holds. For all other ω we define $u(\omega) \equiv 0$. It is easy to check that u is a weakly measurable solution to **(D)**.

(iii) For a pathwise continuous random field $v : [0,T]\times\overline{Q}\times\Omega \to \mathbb{R}$ and $N \in \mathbb{N}$ we define

$$\sigma_N(v) \ := \ \inf\{t > 0 : \sup_{x\in\overline{Q}} |v(t,x)| \geq N\}.$$

We always put $\inf\emptyset = \infty$. Let u_N be the solution to **(D)** corresponding to $(\vartheta, \alpha_N, \beta)$, $N \in \mathbb{N}$, considered in the proof of **(i)**. Then, by setting

$$u(t,x) \ := \ \sum_{N=1}^{\infty} \mathbf{1}_{\{\sigma_{N-1}(u_{N-1})\leq t\leq\sigma_N(u_N)\}} u_N(t,x),$$

$(t,x) \in [0,T] \times \overline{Q}$, with $\sigma_0 \equiv 0$ we get a strongly measurable solution to **(D)**. It is easy to check that this solution is unique. This ends the proof of theorem 1.6

2. A ONE–DIMENSIONAL STOCHASTIC NEUMANN PROBLEM

In the paper of R. Manthey and Ch. Stiewe [9] the following formal stochastic Neumann problem **(N)** is considered.

$$\begin{aligned}
\tfrac{\partial}{\partial t}u(t,x) &= (\Delta u)(t,x) + \alpha(t,x,u(t,x)) + \beta(t,x,u(t,x))\zeta(t,x), \\
&\qquad\qquad\qquad\qquad\qquad\qquad 0 < t \leq T,\ x \in (0,1), \\
u(0,x) &= \vartheta(x), \qquad\qquad\qquad\qquad\qquad\qquad x \in [0,1], \\
\tfrac{\partial}{\partial x}u(t,0) &= \tfrac{\partial}{\partial x}u(t,1) = 0, \qquad\qquad\qquad\qquad t \in (0,T].
\end{aligned}$$

where $\alpha, \beta : [0,T] \times [0,1] \times \mathbb{R} \to \mathbb{R}$ are measurable functions. The symbol ζ stands for a space-time Gaussian white noise on a given complete probability space $(\Omega, \mathcal{F}, \mathbb{P})$.

Let B be a cylindrical Wiener process on $L_2((0,a)^d)$ and suppose that the initial function ϑ is pathwise continuous and independent of B.

Introduce the filtration $\mathbb{F} = (\mathcal{F}_t)_{t\geq 0}$ by

$$\mathcal{F}_t \ := \ \bigcap_{r>t} \sigma\{B_s, \vartheta : s \in [0,r]\},$$

where, in addition, $\mathcal{F}_0$ contains all $\mathbb{P}$-null sets of $\mathcal{F}$.

R. Manthey and Ch. Stiewe [2] use the following notion of a solution to (N).

2.1 DEFINITION. *A random field* $u : [0,T] \times [0,1] \times \Omega \to \mathbb{R}$ *is called solution to* **(N)** *if the following are true.*

(i) $u(t,.)$ *is* $\mathcal{F}_t$*-measurable for any* $t \in [0,T]$.

(ii) *The random field* u *is pathwise continuous on* $[0,T] \times [0,1]$.

(iii) *We have* $u(0,x) = \vartheta(x)$ *and*

$$\begin{aligned} u(t,x) &= \int_0^1 G(t,x,y)\vartheta(y)\,dy \\ &+ \int_0^t \int_0^1 G(t-s,x,y)\alpha(s,y,u(s,y))\,dyds \\ &+ \int_0^t \int_0^1 G(t-s,x,y)\beta(s,y,u(s,y))\,dB_s(y)dy \end{aligned}$$

$\mathbb{P}$*-a.s. for any* $t \in (0,T], x \in [0,1]$.

Here G denotes the Green's function corresponding to $\frac{\partial}{\partial t} - \triangle$ with Neumann boundary conditions. The last term in the last equation is a stochastic integral with respect to B.

We generalize the existence and non-negativity result of the paper of R. Manthey and Ch. Stiewe [9] in the following way.

2.2 THEOREM. *Let* a,b *real numbers with* $a < b$. *Assume that* $\alpha : [0,T] \times [0,1] \times \mathbb{R} \to \mathbb{R}$ *is continuous on* $[0,T] \times [0,1] \times \mathbb{R}$ *and* $\beta : [0,T] \times [0,1] \times \mathbb{R} \to \mathbb{R}$ *is locally Lipschitz continuous in* $u \in \mathbb{R}$. *Furthermore, suppose*

$$\beta(t,x,a) = \beta(t,x,b) = 0$$

and

$$\alpha(t,x,a) \geq 0, \quad \alpha(t,x,b) \leq 0.$$

for all $(t,x) \in [0,T] \times [0,1]$. *Assume*

$$a \leq \vartheta(x) \leq b$$

for all $x \in [0,1]$. *Then there exists at least one solution* u *to* **(N)** *with*

$$a \leq u(t,x) \leq b,$$

$\mathbb{P}$*-a.s. for all* $(t,x) \in [0,T] \times [0,1]$.

3. THE CAUCHY PROBLEM

In this section we consider the formal problem **(C)**

$$\begin{aligned}\frac{\partial}{\partial t}u(t,x) &= (\Delta u)(t,x)+\alpha(t,x,u(t,x))+\beta(t,x,u(t,x))\xi(t,x),\\ &\qquad\qquad\qquad\qquad 0<t\leq T,\ x\in \mathbb{R},\\ u(0,x) &= \vartheta(x), \qquad\qquad\qquad\qquad x\in[0,1],\end{aligned}$$

with $\vartheta(.,\omega)\in C_b(\mathbb{R})$, $\omega\in\Omega$, where $C_b(\mathbb{R})$ denotes the set of all continuous and bounded functions on $\mathbb{R}$.

In view of the theory of ordinary stochastic differential equations, the notion of solution in definition 2.1 is often called "strong solution" (in the probabilistic sense). Now we shall use what is called "weak solution" (in the probabilistic sense).

3.1 DEFINITION. *By a weak solution of* **(C)** *(in the probabilistic sense) we mean a continuous random field* $u:[0,T]\times\mathbb{R}\times\Omega\to\mathbb{R}$ *with the following properties.*

(i) *There exists an* $\mathbb{F}$*-adapted,* $\mathcal{D}'(\mathbb{R})$*-valued standard Wiener process* B.

(ii) $u(t,.)$ *is* $\mathcal{F}_t$*-measurable for any* $t\in[0,T]$.

(iii) u *satisfies* $u(0,x)=\vartheta(x)$, $x\in\mathbb{R}$, *and*

$$\begin{aligned}u(t,x) &= \int_{\mathbb{R}} G(t,x,y)\vartheta(y)\,dy\\ &\quad+\int_0^t\int_{\mathbb{R}} G(t-s,x,y)\alpha(s,y,u(s,y))\,dyds\\ &\quad+\int_0^t\int_{\mathbb{R}} G(t-s,x,y)\beta(s,y,u(s,y))\,dB_s(y)dy\end{aligned}$$

$\mathbb{P}$*-a.s. for any* $t>0, x\in\mathbb{R}$.

The mapping G is now defined by

$$(t,x,y)\ \to\ \frac{1}{\sqrt{4\pi t}}\cdot exp\Big(-(x-y)^2/4t\Big)$$

for $t>0$, $x,y\in\mathbb{R}$. The main result is the following.

3.2 THEOREM. *Let* $\alpha : [0,T] \times \mathbb{R} \times \mathbb{R} \to \mathbb{R}$ *be a continuous mapping ,* $\beta : [0,T] \times \mathbb{R} \times \mathbb{R} \to \mathbb{R}$ *globally Lipschitz continuous, and* $\vartheta : \Omega \times \mathbb{R} \to [a,b]$ *measurable and pathwise continuous.*
Suppose

$$\beta(t,x,a) \;=\; \beta(t,x,b) = 0$$

and

$$\alpha(t,x,a) \;\geq\; 0, \quad \alpha(t,x,b) \leq 0,$$

holds for all $(t,x) \in [0,T] \times \mathbb{R}$. *Then there exists at least one solution* u *to* **(C)** *with*

$$a \;\leq\; u(t,x) \leq b$$

$\mathbb{P}$*-a.s. for all* $(t,x) \in [0,T] \times \mathbb{R}$.

ACKNOWLEDGEMENT

The second author wishes to express her gratitude to the Deutsche Forschungsgemeinschaft for providing financial support.

REFERENCES

1. A. Friedman, *Partial differential equations of parabolic type*, Prentice Hall: Englewood Cliffs, NJ (1964).
2. K. H. Fichtner and R. Manthey, Weak approximations of stochastic equations, *Stochastics and Stochastics Reports*, in print.
3. I.M. Gelfand, N. J. Wilenkin, Verallgemeinerte Funktionen (Distributionen) IV, *Dt. Verlag der Wissenschaften, Berlin*, 1964.
4. D. Henry, *Geometric Theory of Semilinear Parabolic Equations* , Lecture Notes in Mathematics **840**, Springer, New York (1981).
5. P. C. Hohenberg, B. I. Halperin, Theory of dynamical critical phenomena, *Rev. Mod. Phys.* **49.3**, 435–479 (1977).

6. S. Ma., Modern Theory of critical Phenomena, *Benjamin, New York* (1976).

7. R. Manthey, Existence and Uniqueness of a Solution of a Reaction–Diffusion Equation with Polynomial Nonlinearity and White Noise Disturbance, *Math. Nachr.* **128** 121–133 (1986).

8. R. Manthey, On the Cauchy Problem for Reaction–Diffusion Equations with White Noise, *Math. Nachr.* **136** 209–228 (1988).

9. R. Manthey and Ch. Stiewe, Existence and uniqueness of solutions to Volterra's population equation with diffusion and noise, *Stochastics and Stochastics Reports*, **41** no 3 135–161 (1992).

10. R. Manthey and K. Mittmann, Bounded solutions of stochastic reaction–diffusion equations, *to appear in Stochastics and Stochastics Reports* (1993).

11. R. Redlinger, Existence theorems for semilinear parabolic systems with functionals, *Nonlinear Analysis* **8** no 6 667–682 (1984).

ON CLASSICAL SOLUTIONS OF CERTAIN NON-LINEAR INTEGRO-DIFFERENTIAL EQUATIONS

R. MIKULEVIČIUS and H. PRAGARAUSKAS
Institute of Mathematics and Informatics,
Akademijos 4, 2021 Vilnius, Lithuania

INTRODUCTION

We consider the solvability of non-linear convex integro-differential parabolic equations of the second order in the class of functions $C^{2+\alpha}$. The typical example of such equations is the integro-differential Bellman equation arising in the theory of controlled stochastic processes.

The first general results on the solvability in $C^{2+\alpha}$ of the differential Bellman equations with twice differentiable coefficients have been obtained independently by Evans[1,2] and Krylov[3,4]. Remarkable results in the case of Hölder continuous coefficients have been obtained by Safonov[6,7].

In this paper we give a simple extension of Safonov's results given in[7] for integro-differential equations. For simplicity we consider only the Cauchy problem, but similar results hold for boundary value problems.

1. NOTATION AND THE MAIN RESULT

Let $\mathbb{R}^d = (-\infty,\infty)^d$, $d \geqslant 1$; $H_t = (t,1)\times\mathbb{R}^d$. $t \in [0,1]$; $H = H_0$, $\overline{H}_t = [t,1]\times\mathbb{R}^d$. $C(\overline{H}_t) = C^0(\overline{H}_t)$ denotes the space of continuous bounded functions u on $\overline{H}_t$ with the norm

$$|u|_{H_t} = |u|_{0,H_t} = \sup_{(s,x)\in H_t} |u(s,x)|.$$

The partial derivatives of a function u on $\overline{H}$ we shall denote by

$$\partial^l u(t,x) = \frac{\partial^{l_0+l_1+\ldots+l_d}}{\partial t^{l_0}\partial x_1^{l_1}\ldots\partial x_d^{l_d}} u(t,x) = \frac{\partial^{l_0}}{\partial t^{l_0}} u_{x_1^{l_1}\ldots x_d^{l_d}}(t,x),$$

where a multiindex $l = (l_0, l_1, \ldots, l_d) \in \mathbb{N}^{d+1}$, $\mathbb{N} = \{0,1,2,\ldots\}$ with the parabolic order $|l| = 2l_0 + l_1 + \ldots + l_d$, $\partial^0 u = u$.

$C^{n+\alpha}(\overline{H}_t)$ for $n = 0,2$ and $0 \leqslant \alpha < 1$ denotes the space of continuous functions u on $\overline{H}_t$ possessing continuous derivatives $\partial^l u$, $|l| \leqslant n$ with the finite norm

$$\begin{aligned}|u|_{n+\alpha,H_t} = \sum_{|l|\leqslant n} |\partial^l u|_{H_t} + \sum_{|l|=n} \sup_{\substack{(s,x'),(s,x'')\in H_t\\ x'\neq x''}} &|\partial^l u(s,x') \\ &- \partial^l u(s,x'')|/|x'-x''|^\alpha \\ + \sum_{0\leqslant n+\alpha-|l|\leqslant n} \sup_{\substack{(s',x),(s'',x)\in H_t\\ s'\neq s''}} &|\partial^l u(s',x) \\ &- \partial^l u(s'',x)|/|s'-s''|^{(n+\alpha-|l|)/2}.\end{aligned}$$

S_d^+ denotes the set of symmetric non-negative definite $d\times d$- matrices.

Throughout the paper $\alpha \in (0,1)$, $K > 0$ are fixed constants.

Let $G = G(u_{ij}, u_i, u, w, t, x)$ be a function on $\mathbb{R}^{d^2} \times \mathbb{R}^d \times \mathbb{R} \times C^{2+\alpha}(\overline{H}) \times \overline{H}$ with values in $\mathbb{R}$. For $u \in C^{2+\alpha}(\overline{H})$ denote

$$G[u](t,x) = G(u_{x_ix_j}(t,x), u_{x_i}(t,x), u(t,x), u(\cdot,\cdot), t, x).$$

We shall consider the Cauchy problem in the class $C^{2+\alpha}(\overline{H})$:

$$\partial_t u + G[u] = 0 \quad \text{on} \quad H, u(1,\cdot) = 0 \tag{1.1}$$

Let us introduce some classes of functions. Fix arbitrary non-negative constants $\nu \in (0,1)$, K_1, K_2, $(K_\varepsilon)_{\varepsilon\in(0,1)}$ and for brevity denote $\bar{u} = (u_{ij}, u_i, u) \in \mathbb{R}^{d_1}$, where $d_1 = d^2 + d + 1$.

$\mathcal{G}(\alpha, \nu, K, K_1, (K_\varepsilon))$ denotes the set of functions G: $\mathbb{R}^{d_1} \times C^{2+\alpha}(\overline{H}) \times \overline{H} \to \mathbb{R}$ satisfying the following conditions:

a) G is concave with respect to the variable u_{ij},

b) for any $(u_{ij}, u_i, u, w, t, x) \in \mathbb{R}^{d_1} \times C^{2+\alpha}(\overline{H}) \times \overline{H},\ \xi \in \mathbb{R}^d$

$$\nu|\xi|^2 \leqslant G(u_{ij} + \xi_i\xi_j, u_i, u, w, t, x) - G(u_{ij}, u_i, u, w, t, x) \leqslant \nu^{-1}|\xi|^2,$$

c) for any $u_{ij} \in \mathbb{R}^{d^2},\ u_i, u_i' \in \mathbb{R}^d,\ u, u' \in \mathbb{R},\ w \in C^{2+\alpha}(\overline{H}),\ (t,x) \in H$

$$|G(u_{ij}, u_i, u, w, t, x) - G(u_{ij}, u_i', u', w, t, x)| \leqslant K\left(\sum_{i=1}^{d} |u_i - u_i'| + |u - u'|\right),$$

d) for any $\varepsilon \in (0,1),\ t \in [0,1),\ (\bar{u}, w) \in \mathbb{R}^{d_1} \times C^{2+\alpha}(\overline{H})$

$$|G(\bar{u}, w, \cdot)|_{\alpha,H_t} \leqslant K|\bar{u}| + \varepsilon|w|_{2+\alpha,H_t} + K_\varepsilon|w|_{H_t} + K_1,$$

e) if $w,\ w_n \in C^{2+\alpha}(\overline{H})$, $\sup_n |w_n|_{2+\alpha,H} < \infty$ and $w_n \to w$ in $C^2(Q)$ as $n \to \infty$ for any compact subset $Q \subset H$, then $G(\bar{u}, w_n, t, x) \to G(\bar{u}, w, t, x)$ for any $(\bar{u}, t, x) \in \mathbb{R}^{d_1} \times H$.

$\mathcal{G}_1(\alpha, \nu, K, K_1, K_2, (K_\varepsilon))$ denotes the subset of $\mathcal{G}(\alpha, \nu, K, K_1, (K_\varepsilon))$ which consists of all infinitely differentiable functions G with respect to the variable $\bar{u}$ and such that for any $\bar{u} \in \mathbb{R}^{d_1},\ w_1, w_2 \in C^{2+\alpha}(\overline{H}),\ t \in [0,1)$

$$|G(\bar{u}, w_1, \cdot) - G(\bar{u}, w_2, \cdot)|_{H_t} \leqslant K_2|w_1 - w_2|_{H_t}.$$

$\mathcal{G}_2(\alpha, \nu, K, K_1, K_2, (K_\varepsilon))$ denotes the subset of $\mathcal{G}(\alpha, \nu, K, K_1, K_2, (K_\varepsilon))$ which consists of all Frechet differentiable functions G with respect to the variable $(\bar{u}, w)$ and such that the Frechet derivative ∂G satisfies the following conditions:

a) $|\partial G(\bar{u}, w, \cdot)|_{\alpha,H}$ is locally bounded on $\mathbb{R}^{d_1} \times C^{2+\alpha}(\overline{H})$

b) for any $\bar{u}, \bar{u}' \in \mathbb{R}^{d_1},\ w \in C^{2+\alpha}(\overline{H}),\ (t,x) \in H$

$$|\partial G(\bar{u}, w, t, x) - \partial G(\bar{u}', w, t, x)| \leqslant K_2|\bar{u} - \bar{u}'|.$$

Let $\overline{\alpha} = \overline{\alpha}(d, \nu) \in (0,1)$ be the number introduced in Theorem 14.1[7] (in the case of the Dirichlet problem for non-linear elliptic equations of the second order the number $\overline{\alpha}$ is defined in Theorem 4.1[6]).

The main result of the paper is the following theorem.

THEOREM 1.1. *Let $\alpha \in (0, \overline{\alpha}),\ G, G_n \in \mathcal{G}(\alpha, \nu, K, K_1, K_2, (K_\varepsilon))$, $n = 1, 2, \ldots$ and the following assumptions be satisfied:*

a) $G_n \in \mathcal{G}_1(\alpha, \nu, K, K_1, K_2^n, (K_\varepsilon))$, or $G_n \in \mathcal{G}_2(\alpha, \nu, K, K_1, K_2^n, (K_\varepsilon))$.

b) $G_n(\cdot, t, x) \to G(\cdot, t, x)$ as $n \to \infty$ uniformly on bounded subsets of $\mathbb{R}^{d_1} \times C^{2+\alpha}(\overline{H})$ for any $(t,x) \in H$.

Then the problem (1.1) has a solution $u \in C^{2+\alpha}(\overline{H})$ such that

$$|u|_{2+\alpha,H} \leqslant NK_1,$$

where a constant $N = N(\alpha, d, \nu, K, (K_\varepsilon))$.

2. PROOF OF THEOREM 1.1

In order to prove Theorem 1.1 we shall use some auxiliary probabilistic arguments.

Let $\mathbb{C}$ be the set of all continuous functions $\omega : [0,1] \to \mathbb{R}^d$, $\mathcal{F} = \varepsilon(X_s, s \in [0,1])$, $\mathcal{F}_t^s = \varepsilon(X_r, r \in [s,t])$, $\mathbb{F}^s = (\mathcal{F}_{t+}^s)_{t \geqslant s}$, $X_s = X_s(\omega) = \omega_s$, $\omega \in \mathbb{C}$. Let $a: \overline{H} \to S_d^+$ $b: \overline{H} \to \mathbb{R}^d$ be bounded continuous functions and a be non-degenerate.

Define the operator

$$A = A_{t,x} = \sum_{i,j=1}^{d} a_{ij}(t,x) \frac{\partial^2}{\partial x_i \partial x_j} + \sum_{i=1}^{d} b_i(t,x) \frac{\partial}{\partial x_i}.$$

A probability measure $\mathbf{P}$ on $(\mathbb{C}, \mathcal{F})$ is called a solution to the (s,x,A)-martingale problem if $\mathbf{P}(X_r = x,\ 0 \leqslant r \leqslant s) = 1$ and the process

$$u(X_t) - \int_s^t Au(X_r)\, dr$$

is a $(\mathbb{F}^s, \mathbf{P})$-martingale for any $u \in C_0^\infty(\mathbb{R}^d)$.

It is well known that under our assumptions the (s,x,A)-martingale problem has a unique solution for any $(s,x) \in \overline{H}$.

Suppose that $u \in C^2(\overline{H})$, $f \in C(\overline{H})$ and u solves the problem:

$$\partial_t u + Au + f = 0 \qquad \text{on} \quad H,\ u(1,\cdot) = 0. \tag{2.1}$$

then by THE Ito formula

$$u(s,x) = \mathbf{E} \int_s^t f(t, X_t)\, dt, \tag{2.2}$$

where $\mathbf{E}$ denotes the mathematical expectation with respect to the solution of (s,x,A)-martingale problem.

LEMMA 2.1. *Let $0 < \alpha \leqslant \overline{\alpha}$ and $G \in \mathcal{G}_1(\alpha, \nu K, K_1, K_2, (K_\varepsilon))$ or $G \in \mathcal{G}_2(\alpha, \nu K, K_1, K_2, (K_\varepsilon))$. Then there exists $u \in C^{2+\alpha}(\overline{H})$ solving the problem* (1.1).

Proof. 1°. By Theorem15.1[7] there exists a sequence $u^n \in C^{2+\alpha}((\overline{H}))$, $n = 1,2,\ldots$ $(u^\circ = 0)$ solving the problem

$$\partial_t u^n(t,x) + G\Big(u^n_{x_i x_j}(t,x), u^n_{x_i}(t,x), u^{n-1}(t,x), u^{n-1}, t, x\Big) = 0,$$
$$(t,x) \in H, \quad u^n(1,\cdot) = 0 \tag{2.3}$$

From Theorem 14.1[7] it follows that for any $t \in [0,1)$, $n = 1, 2, \dots$

$$|u^n|_{2+\alpha, H_t} \leqslant N\Big(|u^n|_{H_t} + K_1\Big), \tag{2.4}$$

where a constant N depends only on $\alpha, d, \nu, K, (K_\varepsilon)$. In order to estimate $|u^n|_{H_t}$ we write (2.3) in the form:

$$\partial_t u^n + A^n u^n + f^n = 0,$$

where

$$f^n(t,x) = G\big(0, 0, u^{n-1}(t,x), u^{n-1}, t, x\big),$$

$$A^n = \sum_{i,j=1}^{d} a_{ij}^n \frac{\partial^2}{\partial x_i \partial x_j} + \sum_{i=1}^{d} b_i^n \frac{\partial}{\partial x_i}, \tag{2.5}$$

$$a_{ij}^n(t,x) = \int_0^1 G_{u_{ij}}\Big(s u^n_{x_i x_j}(t,x), s u^n_{x_i}(t,x), u^{n-1}(t,x), u^{n-1}, t, x\Big)\, ds$$

and b_i^n are defined by the formula (2.5) with $G_{u_{ij}}$ replaced by G_{u_i}.

By the representation (2.2) for any $(s,x) \in \overline{H}$, $n = 0, 1, \dots$

$$u^n(s,x) = \mathbf{E}^n_{s,x} \int_s^1 f^n(t, X_t)\, dt,$$

where $\mathbf{E}^n_{s,x}$ denotes the mathematical expectation with respect to the solution of the (s, x, A^n)-martingale problem. Now using the definition of the class $\mathcal{G}$ we derive that there exists a constant N such that for any $s \in [0,1)$, $n = 1, 2, \dots$

$$|u^n|_{H_s} \leqslant N \int_s^1 |u^{n-1}|_{2+\alpha, H_t}\, dt + K_1.$$

From this and (2.4) we obtain that there exists a constant N such that for any $s \in [0,1)$, $n = 1, 2, \dots$

$$|u^n|_{2+\alpha, H_s} \leqslant N \int_s^1 |u^{n-1}|_{2+\alpha, H_t}\, dt + N K_1. \tag{2.6}$$

2°. Now we shall prove that

$$\sup_n |u^n|_{2+\alpha,H} < \infty. \tag{2.7}$$

Let $\mu = 1/2N$, where N is a constant from (2.6). Denote for brevity $\rho_t^n = |u^n|_{2+\alpha.H_t}$. Suppose that $c_k = \sup_n \rho_{1-k\mu}^n < \infty$ for some $k \in \mathbb{N}$. Then

$$\rho_t^{n+1} \leqslant N \int_t^{1-k\mu} \rho_s^n \, ds + N'$$

holds for $t \leqslant 1 - k\mu$, where $N' = NK_1 + kc_k/2$. From this it follows that

$$\rho_t^{n+1} \leqslant \frac{1}{2}\rho_t^n + N'$$

and

$$\rho_t^{n+1} \leqslant 2^{-n}\rho_t^1 + N' \sum_n 2^{-n},$$

if $1-(k+1)\mu \leqslant t \leqslant 1 - k\mu$. Therefore $\sup_n \rho_{1-(k+1)\mu}^n < \infty$. Starting from $k = 1$ and noting that $\sup_n \rho_{1-\mu}^n < \infty$ we obtain (2.7).

3°. Denote $v^n = u^{n+1} - u^n$. Obviously v^n solves the problem

$$\begin{aligned} &\partial_t v^n(t,x) + G\Big(u_{x_ix_j}^{n+1}(t,x), u_{x_i}^{n+1}(t,x), u^n(t,x), u^n, t, x\Big) \\ &- G\Big(u_{x_ix_j}^n(t,x), u_{x_i}^n(t,x), u^{n-1}(t,x), u^{n-1}, t, x\Big) = 0, \\ &(t,x) \in H, \quad v^n(1,\cdot) = 0. \end{aligned} \tag{2.8}$$

In order to estimate $|v^n|_H$ we write (2.8) in the form:

$$\partial_t v^n + \overline{A}^n v^n + \bar{f}^n = 0,$$

where

$$\begin{aligned} \bar{f}^n(t,x) = &\, G\Big(u_{x_ix_j}^n(t,x), u_{x_i}^n(t,x), u^n(t,x), u^n, t, x\Big) \\ &- G\Big(u_{x_ix_j}^n(t,x), u_{x_i}^n(t,x), u^{n-1}(t,x), u^{n-1}, t, x\Big), \end{aligned}$$

$$\overline{A}^n = \sum_{i,j=1}^d \bar{a}_{ij}^n \frac{\partial^2}{\partial x_i \partial x_j} + \sum_{i=1}^d \bar{b}_{-i}^n \frac{\partial}{\partial x_i}, \tag{2.9}$$

$$\bar{a}^n_{ij}(t,x) = \int_0^1 G_{u_{ij}}\Big(su^{n+1}_{x_ix_j}(t,x) + (1-s)u^n_{x_ix_j}(t,x), su^{n+1}_{x_i}(t,x)$$

$$+ (1-s)u^n_{x_i}(t,x), u^n(t,x), u^n, t, x\Big)\, ds$$

and in a similar way are defined $\bar{b}^n_i$ replacing $G_{u_{ij}}$ in (2.9) by G_{u_i}.

4°. Let $G \in \mathcal{G}_1$. By the representation (2.2) for any $(s,x) \in H$, $n = 1,2,\ldots$

$$v^n(s,x) = \overline{\mathbf{E}}^n_{s,x} \int_s^1 \bar{f}^n(t, X_t)\, dt, \tag{2.10}$$

where $\overline{\mathbf{E}}^n_{s,x}$ denotes the mathematical expectation with respect to the solution of the $(s,x,\overline{A}^n)$-martingale problem. From this and the definition of the class $\mathcal{G}_1$ we derive, that there exists a constant N such that for any $s \in [0,1]$, $n = 1,2,\ldots$

$$|v^n|_{H_s} \leqslant N \int_s^1 |v^{n-1}|_{H_t}\, dt. \tag{2.11}$$

Now we shall prove that

$$\lim_{n\to\infty} |v^n|_H = 0. \tag{2.12}$$

Let $\mu = 1/2N$, where N is the constant from (2.11). Denote for brevity $\gamma^n_t = |v^n|_{H_t}$. Suppose that $\gamma^n_{1-k\mu} \to 0$ as $n \to \infty$ for some $k \in \mathbb{N}$. Then

$$\gamma^{n+1}_t \leqslant \int_t^{1-k\mu} \gamma^n_s\, ds + \int_{1-k\mu}^1 \gamma^n_s\, ds$$

$$\leqslant 2^{-n}\gamma^1_t + k\mu\left[\sum_{i=l}^{n-1} 2^{-i}\gamma^{n-i}_{1-k\mu} + \sum_{i=0}^{l-1} 2^{-i}\gamma^{n-i}_{1-k\mu}\right]$$

for any $1-(k+1)\mu \leqslant t \leqslant 1-k\mu$ and $l = 1,2,\ldots$. Therefore $\gamma^{n+1}_{1-(k+1)\mu} \to 0$ as $n \to \infty$. Starting from $k = 1$ and noting that $\gamma^n_{1-\mu} \to 0$ as $n \to \infty$ we obtain (2.12).

According to (2.7), (2.12) we can choose a subsequence n_k and a function $u \in C^{2+\alpha}(\overline{H})$ such that $\partial^l u^{n_k} \to \partial^l u$, $\partial^l u^{n_k-1} \to \partial^l u$ as $n_k \to \infty$ uniformly on compact subsets of $\overline{H}$ for any $|l| \leqslant 2$. Passing to the limit in (2.3) as $n_k \to \infty$ we obtain the assertion of the lemma in the case $G \in \mathcal{G}_1$.

Let $G \in \mathcal{G}_2$. According to the theory of linear parabolic equations of the second order (see Chapter IV, Section 14[5]) there exists a constant N such that for any $t \in [0,1)$, $n = 1, \ldots$

$$|v^n|_{2+\alpha,H_t} \leqslant N|\bar{f}^n|_{\alpha,H_t}.$$

From this using the definition of the class $\mathcal{G}_2$ we derive that for each $\varepsilon > 0$ there exists a constant N_ε such that for any $t \in [0,1)$, $n = 2, \ldots$

$$|v^n|_{2+\alpha,H_t} \leqslant \varepsilon|v^{n-1}|_{2+\alpha,H_t} + N_\varepsilon|v^{n-1}|_{H_t}.$$

Further from the representation (2.10) it is easy to derive that there exists a constant N such that for any $s \in [0,1)$, $n = 2, \ldots$

$$|v^{n-1}|_{H_s} \leqslant N \int_s^1 |v^{n-2}|_{2+\alpha,H_t}\, dt.$$

From these estimates we obtain that for each $\varepsilon > 0$ there exists a constant N_ε such that for any $t \in [0,1)$, $n = 3, \ldots$

$$|v^n|_{2+\alpha,H_t} \leqslant \varepsilon|v_{n-1}|_{2+\alpha,H_t} + N_\varepsilon \int_t^1 |v^{n-2}|_{2+\alpha,H_s}.$$

Now in a similar way as (2.12) was derived we obtain that

$$\lim_n |v^n|_{2+\alpha,H} = 0.$$

Passing to the limit in (2.3) as $n \to \infty$ we obtain the assertion of the lemma in the case $G \in \mathcal{G}_2$.

The lemma is proved.

Proof of Theorem 1.1. By Lemma 2.1 for any $n = 1, 2, \ldots$ there exists a function $u^n \in C^{2+\alpha}(\overline{H})$ solving the problem:

$$\partial_t u^n + G_n[u^n] = 0 \quad \text{in} \quad H, \quad u^n(1,\cdot) = 0. \qquad (2.13)$$

From Theorem 14.1[7] it follows that there exists a constant N depending only on $\alpha, d, \nu, K, (K_\varepsilon)$ such that for any $t \in [0,1)$, $n = 1, 2, \dots$

$$|u^n|_{2+\alpha,H_t} \leqslant N\big(|u^n|_{H_t} + K_1\big). \tag{2.14}$$

In order to estimate $|u^n|_{H_t}$ we write (2.13) in the form:

$$\partial_t u^n + \widetilde{A}^n u^n + \tilde{f}^n = 0 \qquad \text{on} \quad H,$$

where

$$\tilde{f}^n(t,x) = G_n\big(0, 0, u^n(t,x), u^n, t, x\big),$$

$$\widetilde{A}^n = \sum_{i,j=1}^{d} \tilde{a}_{ij}^n \frac{\partial^2}{\partial x_i \partial x_j} + \sum_{i=1}^{d} \tilde{b}_i^n \frac{\partial}{\partial x_i}, \tag{2.15}$$

$$\tilde{a}_{ij}^n(t,x) = \int_0^1 G_{nu_{ij}}\Big(su_{x_i x_j}^n(t,x), su_{x_i}^n(t,x), u^n(t,x), u^n, t, x\Big)\, ds$$

and in a similar way is defined $\tilde{b}_i^n$ replacing $G_{nu_{ij}}$ in (2.15) by G_{nu_i}.

By the representation (2.2)

$$u^n(s,x) = \widetilde{E}_{s,x}^n \int_s^1 \tilde{f}^n(t, X_t)\, dt,$$

where $\widetilde{E}_{s,x}^n$ denotes the mathematical expectation with respect to the solution of the $(s, x, \widetilde{A}^n)$- martingale problem. ¿From this it is easy to derive that there exists a constant $N = N(K, (K_\varepsilon))$ such that for any $s \in [0,1)$, $n = 1, 2, \dots$

$$|u^n|_{H_s} \leqslant N \int_s^1 |u^n|_{2+\alpha,H_t}\, dt + K_1.$$

Now using (2.14) and Gronwall's inequality we conclude that

$$\sup_n |u^n|_{2+\alpha,H} \leqslant NK_1,$$

where a constant $N = N(\alpha, \nu, d, K, (K_\varepsilon))$. Therefore the assertion of the theorem follows immediately when passing to the limit in (2.13) for some subsequence n_k.

The theorem is proved.

3. INTEGRO-DIFFERENTIAL BELLMAN EQUATION

In this section we give an example of integro-differential operators arising in the theory of controlled stochastic processes and satisfying the assumptions of Theorem 1.1.

Let $\mathcal{R} = \mathcal{R}(\mathbb{R}^d\backslash\{0\})$ be the space of ε-finite measures on $\mathbb{R}^d\backslash\{0\}$ and $|\mu|$ denotes the variation of an element $\mu \in \mathcal{R}$.

Suppose that for $k \in \mathbb{N}$ we are given measurable functions $a^k: \overline{H} \to S_d^+$, $b^k: \overline{H} \to \mathbb{R}^d$, $r^k: \overline{H} \to \mathbb{R}$, $f^k: \overline{H} \to \mathbb{R}$, $\pi^k: \overline{H} \to \mathcal{R}$, such that for any $(t,x) \in \overline{H}$, $k \in \mathbb{N}$

$$|a^k(t,x)| + |b^k(t,x)| + |r^k(t,x)| + |f^k(t,x)| + \int (|y|^2 \wedge 1)|\pi^k|(t,x,dy) \leqslant K,$$

Let us introduce the operators $P^k_{t,z}$, $(t,z) \in H$, $k \in \mathbb{N}$, acting on functions $u \in C_0^\infty(\mathbb{R}^d)$ by the formula

$$P^k_{t,z}u(x) = \int \left[u(x+y) - u(x) - \sum_{i=1}^d u_{x_i}(x) y_i \mathbf{1}_{\{|y|\leqslant 1\}}\right] \pi^k(t,z,dy) + \sum_{i=1}^d b_i^k(t,z) u_{x_i}(x).$$

Further we shall use the following assumptions.

A1. There exists a constant $\nu > 0$ such that for any $(t,x) \in \overline{H}$, $\xi \in \mathbb{R}^d$, $k \in \mathbb{N}$

$$(a^k(t,x)\xi,\xi) \leqslant \nu|\xi|^2.$$

A2. $a^k_{ij}, r^k, f^k \in C^\alpha(\overline{H})$; $i,j = 1,\dots,d$, $k \in \mathbb{N}$ and

$$|a^k|_{\alpha,H} + |r^k|_{\alpha,H} + |f^k|_{\alpha,H} \leqslant K.$$

A3. For any $u \in C_0^\infty(\mathbb{R}^d)$, $k \in \mathbb{N}$

$$\left|P^k_{\cdot} u(0)\right|_{\alpha,H} \leqslant K|u|_{2,\mathbb{R}^d}.$$

A4. For each $\varepsilon \in (0,1)$, $k \in \mathbb{N}$ there exist measurable non-negative functions $\pi^k_{i,\varepsilon}, \widetilde{\pi}_{i,\varepsilon}: \overline{H} \to \mathcal{R}$, $i = 1,2$ and constants K_ε such that for any $(t,x) \in \overline{H}$

$$|\pi^k|(t,x,dy) \leqslant \pi^k_{1,\varepsilon}(t,x,dy) + \pi^k_{2,\varepsilon}(t,x,dy),$$

$$\int |y|^2 \pi^k_{1,\varepsilon}(t,x,dy) \leqslant \varepsilon, \quad \pi^k_{2,\varepsilon}(t,x,\mathbb{R}^d\backslash\{0\}) \leqslant K_\varepsilon.$$

For $u_{ij} \in \mathbb{R}^{d^2}$, $u \in \mathbb{R}$, $w \in C^{2+\alpha}(\overline{H})$, $(t,x) \in \overline{H}$, $k \in \mathbb{N}$ denote

$$\begin{aligned} L^k(u_{ij},u,w,t,x) &= \sum_{i,j=1}^{d} a^k_{ij}(t,x)u_{ij} + r^k(t,x)u \\ &\quad + P^k_{t,x}w(t,x) + f^k(t,x), \\ G(u_{ij},u,w,t,x) &= \inf_k L^k(u_{ij},u,w,t,x). \end{aligned}$$

THEOREM 3.1. *Let $0 < \alpha \leqslant \overline{\alpha}$ and the assumptions A1–A4 be satisfied. Then the assumptions of Theorem 1.1 hold.*

Proof. For $y = (y_1,\ldots,y_n) \in \mathbb{R}^n$ denote

$$F^n(y) = y_1 \wedge \ldots \wedge y_n.$$

Let $\zeta_n \in C^\infty_0(\mathbb{R}^n)$, $n = 1,2,\ldots$ be non-negative functions such that $\int_{\mathbb{R}^n} \zeta_n(y)\,dy = 1$. For $\varepsilon \in (0,1)$ denote

$$F^{n,\varepsilon}(y) = \varepsilon^{-n} \int_{sym R^n} F^n(z)\zeta_n((y-z)/\varepsilon)\,dz.$$

Let a sequence $\varepsilon_n \downarrow 0$ as $n \uparrow \infty$. Denote

$$G_n(u_{ij},u,w,t,x) = F^{n,\varepsilon_n}\big(L^1(u_{ij},u,w,t,x),\ldots,L^n(u_{ij},u,w,t,x)\big).$$

Noting that $|F^{n,\varepsilon}(y) - F^n(y)| \leqslant \varepsilon$ for any $y \in \mathbb{R}^n$, $\varepsilon > 0$ and $G_n \in G_2$ we obtain the assertion of the theorem.

It is standard in the theory of controlled jump diffusion processes that sufficiently smooth solution of the Bellman equation coincides with the corresponding payoff function. In the rest part of the paper we assume that the assumptions A1–A4 are satisfied for $0 < \alpha \leqslant \overline{\alpha}$ and $\pi^k(t,x,dy)$, $(t,x) \in \overline{H}$, $k \in \mathbb{N}$ are non-negative measures.

Let $\mathcal{B}$ be the set of all measurable functions $\beta\colon \overline{H} \to \mathbb{N}$. For $\beta \in \mathcal{B}$, $(t,x) \in \overline{H}$, denote

$$\mathcal{L}^\beta u(t,x) = \sum_{i,j=1}^{d} a^{\beta(t,x)}_{ij}(t,x)u_{x_ix_j}(t,x) + P^{\beta(t,x)}_{t,x}.$$

Let $\mathbb{S}(s,x,\mathcal{L}^\beta)$ be the set of all solutions to the $(s,x,\mathcal{L}^\beta)$-martingale problem in the Skorokhod space $\mathcal{D}([0,1],\mathbb{R}^d)$. It is well known, that $\mathbb{S}(s,x,\mathcal{L}^\beta)\neq\varnothing$ for any $(s,x)\in\overline{H}$, $\beta\in\mathcal{B}$. Define the payoff function v by the formula

$$v(s,x)=\inf_{\beta\in\mathcal{B},\mathbf{P}\in\mathbb{S}(s,x,\mathcal{L}^\beta)}\mathbf{E}\int_s^1 e^{\varphi^\beta_{s,t}}f^{\beta(t,X_t)}(t,X_t)\,dt,$$

where

$$\varphi^\beta_{s,t}=\int_s^t r^{\beta(u,X_u)}(u,X_u)\,du.$$

THEOREM 3.2. *v is the unique solution in the class $C^{2+\alpha}(\overline{H})$ of the problem*

$$\partial_t u+\inf_{k\in\mathbb{N}}(\mathcal{L}^k u+r^k u+f^k)=0\quad\text{on } H,\qquad u(1,\cdot)=0.\tag{3.1}$$

Proof. By Theorems 1.1 and 2.1 there exists a solution u of the problem (3.1) in the class $C^{2+\alpha}(\overline{H})$. Let $\beta\in\mathcal{B}$, $(s,x)\in\overline{H}$, $\mathbf{P}\in\mathbb{S}(s,x,\mathcal{L}^\beta)$. Then by the Ito formula

$$\begin{aligned}u(s,x)&=-\mathbf{E}\int_s^1 e^{\varphi^\beta_{s,t}}\Big(\partial_t+\mathcal{L}^{\beta(t,X_t)}+r^{\beta(t,X_t)}\Big)u(t,X_t)\,dt\\&\leqslant\mathbf{E}\int_s^1 e^{\varphi^\beta_{s,t}}f^{\beta(t,X_t)}(t,X_t)\,dt\leqslant v(s,x).\end{aligned}$$

Obviously, for any $\varepsilon>0$ there exists $\beta_\varepsilon\in\mathcal{B}$ such that

$$\partial_t u+\mathcal{L}^{\beta_\varepsilon}u+r^{\beta_\varepsilon}+f^{\beta_\varepsilon}\leqslant\varepsilon$$

in $\overline{H}$. By the Ito formula for any $\mathbf{P}\in\mathbb{S}(s,x,\mathcal{L}^{\beta_\varepsilon})$

$$\begin{aligned}u(s,x)&=-\mathbf{E}\int_s^1 e^{\varphi^{\beta_\varepsilon}_{s,t}}\Big(\partial_t+\mathcal{L}^{\beta_\varepsilon(t,X_t)}+r^{\beta_\varepsilon(t,X_t)}\Big)u(t,X_t)\,dt\\&\geqslant\mathbf{E}\int_s^1 e^{\varphi^{\beta_\varepsilon}_{s,t}}f^{\beta_\varepsilon(t,X_t)}(t,X_t)\,dt-\varepsilon\geqslant v(s,x)-\varepsilon.\end{aligned}$$

Therefore $u=v$. The theorem is proved.

REFERENCES

1. L. C. Evans Classical solutions of fully nonlinear convex second order elliptic equations. *Comm. Pure and Appl. Math.* **35** (1982), No 3, 333–363.
2. L. C. Evans (1983) Classical solutions of the Hamilton–Jacobi–Bellman equation for uniformly elliptic operators. *Trans. Amer. Math. Soc.* **275** (1983), No 1, 245–255.
3. N. V. Krylov Restricted inhomogeneous elliptic and parabolic equations. *Math. USSR Izvestija* **46** (1982), 487–523 (in Russian).
4. N. V. Krylov Restricted inhomogeneous elliptic and parabolic equations in a domain. *Math. USSR Izvestija* **47** (1983), 75–108 (in Russian).
5. O. A. Ladyzhenskaja, V. A. Solonnikov, N. N. Uraltseva *Linear and Quasi — Linear Equations of Parabolic Type.* Transl. Math. Monographs, vol. 23, Amer. Math. Soc., Providence, R.I. (1968).
6. M. V. Safonov On classical solution of non-linear elliptic equations of the second order. *Math. USSR Izvestija* **52** (1988), 1272–1287 (in Russian).
7. M. V. Safonov *On boundary value problems for non-linear elliptic and parabolic equations of the second order.* ScD, Thesis, LOMI, Leningrad (in Russian). (1988).

ON THE DECOMPOSITION OF STRONG MARKOV CONTINUOUS SEMIMARTINGALES

WOLFGANG SCHMIDT

Friedrich–Schiller–Universität Jena, Institut für Stochastik, UHH 17.OG, O–6900 Jena, Germany

Abstract Let X be a continuous strong Markov process which is also a semimartingale. We derive an explicit formula for the bounded variation part in the semimartingale decomposition of X. The bounded variation part involves the local time and certain occupation times of X.

1. INTRODUCTION

Let X be a continuous strong Markov process on the real axis $\mathbf{R}$. Supposing that X is also a semimartingale we are interested in the explicit form of the semimartingale decomposition

$$X = X_0 + M + V,$$

where M is a continuous local martingale and V is locally of bounded variation.

The study of the relationships between Markov processes and martingales has a rather long history (cf. for example [3], [4], [5], [6], [8], [12]).

In their paper [3] CINLAR and JACOD proved that every semimartingale Hunt process is a time change of a solution to a stochastic differential equation in the sense of K. Ito. However the time change and the coefficients of the stochastic differential equation are rather arbitrary (cf. [9]).

We are interested in the semimartingale decomposition of X itself. S. MELEARD proved in [8] that a regular continuous strong Markov process on $[0,1]$ with the identity as scale is a semimartingale which admits the decomposition

$$X_t = X_0 + M_t + L^X(t,0) - L^X(t,1),\ t \geq 0,$$

where L^X is the symmetric local time of X.

For a regular continuous strong Markov semimartingale on the whole axis $\mathbf{R}$ it was shown in [9] that the bounded variation part V in the decomposition of X is given by

$$V_t = 1/2 \int L^X(t,y)(\bar{g}(y))^{-1} dg(y), \qquad (1)$$

where g is a right continuous function of locally bounded variation related to the scale of X and $\bar{g}(y) = 1/2(g(x) + g(x-))$.

In [2] CINLAR proved a nice result on the structure of continuous strong Markov processes X on the real axis having paths of locally bounded variation. Such processes are simply deterministic functions of their initial value and time:

$$X_t = h(X_0, t),\ t \geq 0, \text{ a.s.}$$

Generalizing the above results, in this paper we show that the bounded variation part of an arbitrary continuous strong Markov semimartingale X consists of a sum of the local times of X at certain boundary points, an integral of the local time and two deterministic functions of the occupation time of X in two specific subsets of the so called singular points of X. To this end we will refine the well-known classification of points of the state interval of X. An important tool for proving the result is also the notion of a weakly additive functional introduced in [10].

2. NOTATION

Let I be an interval of the real axis $\mathbf{R}$. Denote by $\mathcal{B}(I)$ the σ-algebra of the Borel sets from I and by $\mathcal{B}^u(I)$ its universal completion. Let us be given a family $(\Omega, \mathcal{F}^o, \mathbf{P}_x, x \in I)$ of probability spaces. We suppose that $\mathbf{P}_x(A)$ is universally measurable in x for every $A \in \mathcal{F}^o$. For a probablity μ on $(I, \mathcal{B}^u(I))$ define

$$\mathbf{P}_\mu(A) = \int_A \mathbf{P}_x(A)\mu(dx),\ A \in \mathcal{F}^o,$$

and let $\mathcal{F}^\mu$ be the $\mathbf{P}_\mu$-completion of $\mathcal{F}^o$. We set $\mathcal{F} = \bigcap \mathcal{F}^\mu$ where the intersection is taken over all probabilities μ on $(I, \mathcal{B}^u(I))$. In the sequel *a.s.* means $\mathbf{P}_x$-a.s. for every $x \in I$.

Now let $X = (X_t)_{t\geq 0}$ defined on $(\Omega, \mathcal{F}, \mathbf{P}_x, x \in I)$ be a continuous stochastic process taking values in I together with a filtration $\mathbf{F}= (\mathcal{F}_t)_{t\geq 0}$ of sub-σ-algebras of $\mathcal{F}$. We suppose that X is $\mathbf{F}$-adapted and that $\mathcal{F}_0$ is complete in $\mathcal{F}$ w.r.t. the family $(\mathbf{P}_\mu)$ (cf. [1]). Define $\mathcal{F}_t^o = \sigma\{X_s : s \leq t\}$, $0 \leq t < \infty$, and let $\mathcal{F}_t^X$ denote the completion of $\mathcal{F}_t^o$ in $\mathcal{F}$ w.r.t. the family $(\mathbf{P}_\mu)$. We write $\mathbf{F}^X = (\mathcal{F}_t^X)_{t\geq 0}$.

Definition. We say that $(X, \mathbf{F})$ defined on $(\Omega, \mathcal{F}, \mathbf{P}_x, x \in I)$ is a *continuous strong Markov process*, if

(i) $\mathbf{P}_x(X_0 = x) = 1,\ x \in I$,

(ii) X admits shift operators $\theta = (\theta_t)_{t\geq 0}$, i.e. θ is a semigroup of operators $\theta_t \mid \Omega \to \Omega$ such that
$X_{t+s} = X_s \circ \theta_t,\ t \geq 0,\ s \geq 0,$ a.s.,

(iii) for every nonnegative $\mathcal{F}_\infty^X$–measurable random variable Y
$\mathbf{E}_x(Y \circ \theta_t \mid \mathcal{F}_t) = \mathbf{E}_{X_t}(Y)$ $\mathbf{P}_x$-a.s.,

(iv) for every $\mathbf{F}_+^X$ stopping time T and every nonnegative $\mathcal{F}_\infty^X$-measurable random variable Y
$\mathbf{E}_x(Y \circ \theta_T \mathbf{1}_{\{T<\infty\}} \mid \mathcal{F}_{T+}^X) = \mathbf{E}_{X_T}(Y)\mathbf{1}_{\{T<\infty\}}$ $\mathbf{P}_x$-a.s.

It is known that for a continuous strong Markov process X we have $\mathbf{F}_+^X = \mathbf{F}^X$ (cf. [1], I (8.12)).

In the following we always suppose that $(X, \mathbf{F})$ is a continuous strong Markov process.

If $(X, \mathbf{F})$ is moreover a semimartingale w.r.t. every $\mathbf{P}_x$, $x \in I$, we call $(X, \mathbf{F})$ a *continuous strong Markov semimartingale*. Then it was shown by CINLAR, JACOD, PROTTER and SHARPE [4] that there exists a "universal" decomposition of X, i.e.

$$X = X_0 + M(X) + V(X) \quad \text{a.s.},$$

where $(M(X), \mathbf{F}^X)$ is a continuous local martingale w.r.t. every $\mathbf{P}_x$, $x \in I$, $(V(X), \mathbf{F}^X)$ has paths with bounded variation on finite intervals a.s., and $M_0(X) = V_0(X) = 0$. (cf. [4], Theorem (3.12), [13]).

The goal of this paper is to clarify the structure of the process $V(X)$. To this purpose we have to introduce some further notation.

Define the *right local time* $L_+^X(.,y)$ of X in the point $y \in I$ by

$$L_+^X(t,y) = \mid X_t - y \mid - \mid X_0 - y \mid - \int_0^t sign_-(X_s - y)dX_s,$$

where $sign_- = -\mathbf{1}_{(-\infty,0]} + \mathbf{1}_{(0,\infty)}$ and the stochastic integral is taken such that it is a version of the corresponding $\mathbf{P}_x$–stochastic integral for every $x \in I$ (cf. [4]). Analogously the *left local time* L_-^X is defined replacing $sign_-$ by $sign_+ = -\mathbf{1}_{(-\infty,0)} + \mathbf{1}_{[0,\infty)}$. Finally let the (symmetric) *local time* of X be defined by

$$L^X(t,y) = 1/2\left(L_+^X(t,y) + L_-^X(t,y)\right).$$

Let us recall the classification of points of the state interval I for the continuous strong Markov process X (cf. [7]). For $A \in \mathcal{B}(I)$ let

$$D_A = \inf\{t \geq 0 : X_t \in A\}$$

and abbreviate

$$D_x = D_{\{x\}},\, D_{x+} = D_{(x,\infty)\cap I},\, D_{x-} = D_{(-\infty,x)\cap I}.$$

Then D_x, D_{x+}, D_{x-} are $\mathbf{F}^X$–stopping times and by Blumenthal's 0-1 law $\mathbf{P}_x(D_{x\pm} = 0) \in \{0,1\}$. The point $x \in I$ is called

$$\begin{array}{rl} regular & \text{if } \mathbf{P}_x(D_{x+} = 0) = \mathbf{P}_x(D_{x-} = 0) = 1, \\ left\ singular & \text{if } \mathbf{P}_x(D_{x+} = 0) = 0, \\ right\ singular & \text{if } \mathbf{P}_x(D_{x-} = 0) = 0, \\ absorbing & \text{if } \mathbf{P}_x(D_{x+} = 0) = \mathbf{P}_x(D_{x-} = 0) = 0. \end{array}$$

The set of all regular points is denoted by R whereas we write K_+ (resp. K_-) for the set of all right (resp. left) singular points. Finally $E = K_+ \cap K_-$ is the set of absorbing points. The set R is open and thus a union of disjoint open intervals:

$$R = \bigcup_n (a_n, b_n). \tag{2}$$

The intervals (a_n, b_n) are called the components of R.

The process X is called *regular* if every inner point of I is regular. It is well-known that there exists a function $p \mid R \to \mathbf{R}$ which is increasing and continuous on each component (a,b) of R and satisfies

$$\mathbf{P}_x(D_y < D_z) = (p(z) - p(x))/(p(z) - p(y))$$

for every $a < y < x < z < b$. The function p is called a *scale* for X.

For what follows we need a refined decomposition of the set of the singular points. To this end we define the sets

$$S_+ = K_+ \setminus (\{a_n : a_n \in K_+\} \cup E) \tag{3}$$
$$S_- = K_- \setminus (\{b_n : b_n \in K_-\} \cup E). \tag{4}$$

3. THE SEMIMARTINGALE DECOMPOSITION

Here is the main result of this paper.

Theorem. *Let $(X, \mathbf{F})$ defined on $(\Omega, \mathcal{F}, \mathbf{P}_x, x \in I)$ be a continuous strong Markov semimartingale with the semimartingale decomposition*

$$X = X_0 + M(X) + V(X).$$

For a scale $p \mid R \to \mathbf{R}$ of X denote by q_n the inverse of p restricted to (a_n, b_n):

$$q_n \mid (p(a_n+), p(b_n-)) \to (a_n, b_n).$$

Then q_n is a difference of convex functions and we define $g \mid R \to [0, \infty)$ by

$$g(x) = (q_n)'_+(p(x)),\ x \in (a_n, b_n),\ n = 1, 2, \ldots.$$

There exist $\mathcal{B}^u(I) \otimes \mathcal{B}(\mathbf{R}_+)$-measurable functions

$$d_+ \mid I \times \mathbf{R}_+ \to [0, \infty),\ d_- \mid I \times \mathbf{R}_+ \to [0, \infty)$$

which are continuously increasing in the second variable such that

$$\begin{aligned} V_t(X) &= 1/2 \int_R L^X(t, y)(\bar{g}(y))^{-1} dg(y) \\ &\quad + \sum_{n:\, a_n \in K_+} L^X(t, a_n) - \sum_{n:\, b_n \in K_-} L^X(t, b_n) \\ &\quad + d_+\left(X_0, \int_0^t \mathbf{1}_{S_+}(X_s) ds\right) - d_-\left(X_0, \int_0^t \mathbf{1}_{S_-}(X_s) ds\right), \end{aligned}$$

$t \geq 0$, a.s., where $\bar{g}(x) = 1/2\,(g(x) + g(x+))$, $x \in R$.

Corollary (cf. [2]). *Let $(X, \mathbf{F})$ on $(\Omega, \mathcal{F}, \mathbf{P}_x, x \in I)$ be a continuous strong Markov process admitting increasing paths a.s. Then there exists a $\mathcal{B}^u(I) \otimes \mathcal{B}(\mathbf{R}_+)$-measurable function $h \mid I \times \mathbf{R}_+ \to I$ such that*

$$X_t = h(X_0, t),\ t \geq 0,\ a.s.$$

The function h satisfies $h(h(x,t), s) = h(x, t+s)$, $x \in I$, $s \geq 0$ and $h(x, .)$ is continuous and increasing.

The assertion of the corollary follows from the theorem since $S_+ = I \setminus \{\sup I\}$ under the conditions of the corollary. In fact the result of the corollary was already shown by CINLAR in [2] and we will make substantially use of it in the proof of the theorem.

However, a detailed proof of the theorem is rather long (cf. [11]). We will therefore restrict ourselves to a sketch of the proof. Demonstrating the main ideas will take the reader in the position to fill the gaps by himself.

Proof of the Theorem (sketch). We investigate the behaviour of $V(X)$ "on" the sets R, $\{a_n\}$ for $a_n \in K_+$ (resp. $\{b_n\}$ for $b_n \in K_-$) and S_+ (resp. S_-).

First we show that

$$\int_0^t \mathbf{1}_R(X_s) dV_s(X) = 1/2 \int_R L^X(t,y)(\bar{g}(y))^{-1} dg(y),\ t \geq 0, \text{ a.s.}$$

For this it suffices to prove that for a component (a, b) of R and (c, d) with $[c, d] \subset (a, b)$ it holds

$$\int_0^t \mathbf{1}_{(c,d)}(X_s) dV_s(X) = 1/2 \int_{(c,d)} L^X(t,y)(\bar{g}(y))^{-1} dg(y),\ t \geq 0, \text{ a.s.}$$

As in [8] it can be shown that $Y_t = p((X_t \vee c) \wedge d)$, $t \geq 0$, is a semimartingale with decomposition

$$Y_t = Y_0 + M_t + L^Y(t,c) - L^Y(t,d),\ t \geq 0.$$

If q denotes the inverse of p restricted to (a, b) then the process $q(Y) = (X \vee c) \wedge d$ is a semimartingale. The function q is therefore a difference of convex functions on $[p(c), p(d)]$. The semimartingale decomposition of $(X \vee c) \wedge d$ can be calculated using the generalized Ito's formula

$$(X_t \vee c) \wedge d = (X_0 \vee c) \wedge d + \int_0^t q'(Y_s) dY_s + 1/2 \int L^Y(t,y) dq'_+(y).^*$$

$^*q'$ denotes the symmetric derivative of q, i.e. $q' = 1/2(q'_+ + q'_-)$.

This yields

$$\int_0^t \mathbf{1}_{(c,d)}(X_s)dV_s(X) = 1/2 \int_{(p(c),p(d))} L^Y(t,y)dq'_+(y),$$

and using the relation $L^X(.,q(y)) = L^Y(.,y)q'(y)$, $y \in (c,d)$, (cf. [11]) the assertion follows.

Now for $a_n \in K_+$ consider the process

$$k_t = \int_0^t \mathbf{1}_{\{a_n\}}(X_s)dV_s(X),\ t \geq 0.$$

We will show that $k_t = L^X(t,a_n)$, $t \geq 0$, a.s. For $t \leq D_{a_n}$ we have obviously $k_t = 0 = L^X(t,a_n)$. Using the Markov property of X and the additivity of $L^X(.,a_n)$ and k it remains to verify that $\mathbf{P}_{a_n}\left(k_t = L^X(t,a_n)\right) = 1$, $t \geq 0$. In view of $a_n \in K_+$ we have $X = (X - a_n)^+ + a_n$, $\mathbf{P}_{a_n}$-a.s., and applying the generalized Ito's formula yields $k_t = L^X(t,a_n)$, $t \geq 0$, $\mathbf{P}_{a_n}$-a.s.

The most interesting part of the proof is the analysis of the behaviour of $V(X)$ "on" S_+ (resp. S_-).

Lemma. *For every $x \in I$ there exist continuous functions*

$$d_+(x,.) \mid \mathbf{R}_+ \to [0,\infty),\ d_-(x,.) \mid \mathbf{R}_+ \to [0,\infty)$$

such that

$$\int_0^t \mathbf{1}_{S_+}(X_s)dV_s(X) = d_+\left(x, \int_0^t \mathbf{1}_{S_+}(X_s)ds\right),\ t \geq 0,\ \mathbf{P}_x\text{-a.s.} \quad (5)$$

$$\int_0^t \mathbf{1}_{S_-}(X_s)dV_s(X) = -d_-\left(x, \int_0^t \mathbf{1}_{S_-}(X_s)ds\right),\ t \geq 0,\ \mathbf{P}_x\text{-a.s.} \quad (6)$$

Proof of the Lemma. We restrict ourselves to the proof of the statement concerning the behaviour of $V(X)$ "on" S_+. The main idea of the proof consists in the following. By a suitable time change we transform X into a continuous strong Markov process Y admitting increasing path which behave "on S_+ like the paths of X". Then by CINLAR's result [2] (cf. the Corollary) Y_t is a deterministic function of Y_0 and t. Transforming this back we obtain "how the paths of X behave on S_+".

Clearly a time change transforming X into an increasing process is in general discontinuous. Our time change will be the right inverse of a so called weakly additive functional of X, cf. [10].

To begin with, we first observe that it suffices to prove that for every $x \in S_+$ there exists a continuous increasing function $d_+(x,.)$ such that (5) is fulfilled.

Indeed, suppose we have shown (5) for every $x \in S_+$. Then for $x \in I$ let $s_+(x) = \inf\{y \geq x : y \in S_+\}$ and define

$$d_+(x,.) = \begin{cases} d_+(s_+(x),.) & , \quad s_+(x) \in S_+ \\ 0 & , \quad s_+(x) \in E \cup \{\infty\}. \end{cases}$$

Then $D_{S_+} = D_{s_+(x)}$ $\mathbf{P}_x$–a.s. on $\{D_{S_+} < \infty\}$, $s_+(x) \in S_+$ if $\mathbf{P}_x(D_{S_+} < \infty) > 0$ and using the strong Markov property it is easy to show that (5) holds for every $x \in I$.

Now let $x \in S_+$ be fixed and set $d = \inf\{y \geq x : y \in K_-\} \wedge \sup\{y \geq x : \mathbf{P}_x(D_y < \infty) > 0\}$. Then $d > x$ and we have $\mathbf{P}_x$–a.s.

$$\begin{aligned} \int_0^t \mathbf{1}_{S_+}(X_s) dV_s(X) &= \int_0^{t \wedge D_d} \mathbf{1}_{S_+}(X_s) dV_s(X) \\ \int_0^t \mathbf{1}_{S_+}(X_s) ds &= \int_0^{t \wedge D_d} \mathbf{1}_{S_+}(X_s) ds. \end{aligned}$$

Therefore without restricting the generality we can assume that $I = [x, d]$ with $E = \{d\}$ or $I = [x, d)$ if d is not in the initial state interval of X, and

$$\mathbf{P}_x(D_y < \infty) = 1 \text{ for every } y < d.$$

This yields $[x, d) \cap K_- = \emptyset$ and

$$[x, d) = S_+ \cup \bigcup_n [a_n, b_n), \; a_n, b_n \in K_+, \; R = \bigcup_n (a_n, b_n).$$

Now denote $V_t^+ = \int_0^t \mathbf{1}_{S_+}(X_s) dV_s(X), t \geq 0$. Then

$$V_t^+ = \ell([X_0, X_t] \cap S_+), \; t \geq 0, \text{ a.s.}^\dagger \tag{7}$$

$^\dagger \ell$ denotes the Lebesgue measure

In fact, one can easily prove that $\int_0^t \mathbf{1}_{S_+}(X_s)dM_s(X) = 0$ (cf. Remark 2 below) and thus

$$\begin{aligned} V_t^+ &= \int_0^t \mathbf{1}_{S_+}(X_s)dX_s \\ &= X_t - X_0 - \int_0^t \mathbf{1}_{\bigcup_n [a_n,b_n)}(X_s)dX_s \\ &= X_t - X_0 - \int_0^t \sum_n \mathbf{1}_{[D_{a_n},D_{b_n})}(s)dX_s \\ &= X_t - X_0 - \ell\left([X_0, X_t] \cap \bigcup_n [a_n, b_n) \right) \\ &= \ell\left([X_0, X_t] \cap S_+\right), \, t \geq 0, \, \mathbf{P}_x\text{-a.s.} \end{aligned}$$

Now let $L \mid [x,d) \to [0,\infty)$ be a continuous strictly increasing function with $\sum_n (L(b_n) - L(a_n)) = \infty$ if $\sup_n b_n = d$. We define

$$\begin{aligned} A_t &= \int_{[X_0, \max_{s \leq t} X_s] \cap \bigcup_n [a_n,b_n)} L(dy) + \int_0^t \mathbf{1}_{S_+}(X_s)ds, \\ T_t &= \inf\{s \geq 0 : A_s > t\}, \, t \geq 0. \end{aligned}$$

Then $A_\infty = \infty$ and A is a weakly additive functional of $(X, \mathbf{F})$ (cf. [10]). Then by the Theorem in [10] $Y_t = X_{T_t}, t \geq 0$, is a continuous strong Markov process on $(\Omega, \mathcal{F}, \mathbf{P}_x, x \in I)$ with $Y_0 = X_0$ a.s. The definition of A yields that every $y \in I$ is right singular for Y. Consequently Y has increasing paths and by CINLAR's result [2] there exists a function $h \mid I \times \mathbf{R}_+ \to I$ such that $Y_t = h(X_0, t)$, $t \geq 0$, a.s. and $h(y,.)$ is continuous and strictly increasing. Now (7) implies

$$V_{T_t}^+ = \ell\left([x, h(x,t)] \cap S_+\right), \, t \geq 0, \, \mathbf{P}_x\text{–a.s.} \tag{8}$$

Next we show that there exists a function $d_+(x,.) \mid \mathbf{R}_+ \to [0,\infty)$ such that

$$V_{T_t}^+ = d_+\left(x, \int_0^t \mathbf{1}_{S_+}(Y_s)ds\right), \, t \geq 0, \, \mathbf{P}_x\text{–a.s.} \tag{9}$$

To this end let $g(x,.) \mid [x, h(x,\infty)) \to [0,\infty)$ be the inverse of $h(x,.)$. We set $g^*(x,t) = \int_0^t \mathbf{1}_{S_+}(s)g(x,ds)$, $x \leq t < h(x,\infty)$. Then

$$g^*(x,t) = \int_0^{g(x,t)} \mathbf{1}_{S_+}(Y_s)ds, \, \mathbf{P}_x\text{–a.s.}$$

and therefore

$$g^*(x,h(x,t)) = \int_0^t \mathbf{1}_{S_+}(Y_s)ds,\ t \geq 0,\ \mathbf{P}_x\text{-a.s.} \tag{10}$$

Now let $h^*(x,.)$ be the right inverse of $g^*(x,.)$. We have

$$h^*(x,g^*(x,t)) = \begin{cases} t & ,\quad t \in S_+ \\ b_n & ,\quad t \in [a_n,b_n),\ x \leq t < h(x,\infty) \end{cases}$$

and consequently

$$\ell([x,h^*(x,g^*(x,t))] \cap S_+) = \ell([x,t] \cap S_+),\ x \leq t < h(x,\infty).$$

Then (8), (10) imply

$$\begin{aligned} V^+_{T_t} &= \ell([x,h(x,t)] \cap S_+) \\ &= \ell([x,h^*(x,g^*(x,h(x,t)))] \cap S_+) \\ &= \ell\left(\left[x,h^*\left(x,\int_0^t \mathbf{1}_{S_+}(Y_s)ds\right)\right] \cap S_+\right),\ t \geq 0,\ \mathbf{P}_x\text{–a.s} \end{aligned}$$

Now defining

$$d_+(x,t) = \begin{cases} \ell([x,h^*(x,t)] \cap S_+), & 0 \leq t < \int_0^\infty \mathbf{1}_{S_+}(h(x,s))d \\ \lim\limits_{v \uparrow \int_0^\infty \mathbf{1}_{S_+}(h(x,s))ds} \ell([x,h^*(x,v)] \cap S_+), & \infty > t \geq \int_0^\infty \mathbf{1}_{S_+}(h(x,s)) \end{cases}$$

yields (9). To prove the desired formula for V^+ we change the time in (9) by the right inverse A of T,

$$V^+_{T_{A_t}} = d_+\left(x, \int_0^{A_t} \mathbf{1}_{S_+}(Y_s)ds\right).$$

Using the definition of A it is easy to show that

$$\int_0^{A_t} \mathbf{1}_{S_+}(Y_s)ds = \int_0^t \mathbf{1}_{S_+}(X_s)ds.$$

On the other hand, we have $V^+_{T_{A_t}} = V^+_t$, $t \geq 0$, $\mathbf{P}_x$–a.s., since V^+ is constant on each interval where A is constant. This ends the proof of the lemma. □

Now summarizing our considerations and taking into account that

$$\int_0^t \mathbf{1}_E(X_s)dV_s(X) = 0,\ t \geq 0, \text{ a.s.}$$

finishes the sketch of the proof of the theorem. □

Remarks.

1. The functions d_+, d_- in the assertion of the theorem are uniquely determined by the process X. Moreover it can be shown (cf. [11]) that d_+, d_- are explicitly connected with the process X by

$$d_+(x,t) = \ell\left([x, h_+^*(x,t)] \cap S_+\right),\ 0 \leq t \leq f_+(x,\infty),$$

where

$$\begin{aligned} h_+^*(x,t) &= \inf\{z \geq x : f_+(x,z) \geq t\}, 0 \leq t \leq f_+(x,\infty), \\ f_+(x,z) &= \mathbf{E}_x \int_0^{D_z} \mathbf{1}_{S_+}(X_s)ds / \mathbf{P}_x(D_{S_+} < \infty),\ z \geq x \end{aligned}$$

and $0/0 = 0$. An analogous formula holds for the function d_-.

2. Concerning the martingale part in the decomposition of X it seems to be hard to get an explicit description as for the bounded variation part. However it is straightforward to verify that

$$M_t(X) = \int_0^t \mathbf{1}_R(X_s)dM_s(X),\ t \geq 0, \text{ a.s.}$$

Indeed, from the occupation time formula $\int_0^t \mathbf{1}_{K_+}(X_s)d\langle M(X)\rangle_s = \int_{K_+} L_-^X(t,y)dy = 0,\ t \geq 0$, since $L_-^X(.,y) = 0$ for $y \in K_-$, and analogously for K_-.

3. In order to describe continuous strong Markov processes by stochastic differential equations the assertion of the theorem indicates that one has to study stochastic differential equations with "generalized" drift terms as in the theorem.

REFERENCES

1. R.M. Blumenthal, R.K. Getoor. *Markov Processes and Potential Theory.* Academic Press, 1968
2. E. Cinlar. Markov additive processes I, II. *Z. Wahrscheinlichkeitstheorie verw. Gebiete,* **24** 85–93, 94–121 (1972)
3. E. Cinlar, J. Jacod. Representation of semimartingale Markov processes in terms of Wiener processes and Poisson random measures. In *Seminar on Stochastic Processes*, Progress in Probability and Statistics **1**, 159–242, Birkhäuser, 1981
4. E. Cinlar, J. Jacod, P. Protter, M.J. Sharpe. Semimartingales and Markov processes. *Z. Wahrscheinlichkeitstheorie verw. Gebiete* **54**, 161–219 (1980)
5. H.J. Engelbert, W. Schmidt. Strong Markov continuous local martingales and solutions of one–dimensional stochastic differential equations I. *Math. Nachr.* **143**, 167–184 (1989)
6. H.J. Engelbert, W. Schmidt. Strong Markov continuous local martingales and solutions of one–dimensional stochastic differential equations II. *Math. Nachr.* **144**, 241–281 (1989)
7. K. Ito, H.P. McKean. *Diffusion Processes and Their Sample Paths.* Berlin: Springer 1965
8. S. Meleard. Application du calcul stochastique à l'etude de processes de Markov réguliers sur $[0,1]$. *Stochastics* **19**, 41–82 (1986)
9. W. Schmidt. On semimartingale diffusions and stochastic differential equations. *Stochastics and Stochastics Reports* **29**, 407–424 (1990)
10. W. Schmidt. Weakly additive functionals and time changes of strong Markov processes. In *Stochastic Processes and Related Topics*, Volume **61** of *Mathematical Research*, pp. 145–152, Berlin: Akademie–Verlag 1991
11. W. Schmidt. *Über streng Markowsche stetige Semimartingale.* Habilitationsschrift, Friedrich–Schiller–Universität Jena 1992
12. A.V. Skorohod. On homogeneous continuous Markov processes that are martingales. *Theory Probab. Appl.* **8**, 355–365 (1963)
13. C. Stricker, M. Yor. Calcul stochastique dépendant d'un paramètre. *Z. Wahrscheinlichkeitstheorie verw. Gebiete* **45**, 109–133 (1978)

SOME REMARKS ON HEDGING OF CONTINGENT CLAIMS

CHRISTOPHE STRICKER

Université de Franche-Comté
Laboratoire de Mathématiques, URA CNRS 741
16, Route de Gray
25030 Besançon cedex

Abstract This is a survey on some recent results on pricing of contingent claims in an incomplete market. The problem of pricing contingent claims from the price dynamics of some securities is well understood in the context of a complete financial market. In order to avoid any arbitrage opportunity, we assume the existence of a martingale measure which is unique in a complete market. Then the price of a contingent claim is its expectation with respect to this unique martingale measure. On the other hand when the market is incomplete there are several martingale measures and therefore several prices for a general contingent claim. In this paper we study the relationship between the maximum price and the existence of a hedging strategy, and give necessary and sufficient conditions for a contingent claim to be representable with respect to the discounted price process.

1. INTRODUCTION

The "Fundamental Theorem of Asset Pricing" states that the existence of an equivalent martingale measure for the discounted price process S is "essentially" equivalent to the absence of arbitrage opportunities. The theorem is rightly termed "fundamental" as it allows to relate the concept of pricing by arbitrage - which has experienced increasing importance since the seminal papers of F. Black and M.

Scholes (1973) and R. Merton (1973) - with the machinery of martingale theory. When the market is complete, then J.M. Harrison and S.R. Pliska (1981) showed that the collection of martingale measures is a singleton (see the right definition of completeness in Chatelain and Stricker (1992)). In this case the price of a contingent claim is the expected value of this claim with respect to the unique martingale measure. However, the price system may not suffice in itself to give a complete description of the exogenous uncertain environment ; other information, either inside or outside the market, may be available, and might influence market fluctuations. Therefore, it is of interest to let the information structure be as general as possible, and thus place oneself in an "incomplete" market. As mentionned at the beginning the "absence of arbitrage opportunities", which has a natural economic interpretation, implies the existence of a martingale measure. Since the market is not complete, there are several martingale measures and there is a price range for the actual market price of a general contingent claim. We study the relationship between the maximum price and the existence of a hedging strategy, and give necessary and sufficient conditions for a contingent claim to be representable with respect to the discounted price process.

Let $(\Omega, \mathcal{F}, (\mathcal{F}_t)_{0\leq t\leq 1}, P)$ be a filtered probability space satisfying the usual conditions. We assume $\mathcal{F} = \mathcal{F}_1$ and $\mathcal{F}_0$ is P-trivial. We denote by $H^*.X$ the stochastic integral of the predictable process H with respect to the vector valued semimartingale $X := (X^1, \ldots, X^d)$ (see J. Jacod (1979) for precise definitions). Let us make some other definitions before stating the main results. From now on we suppose that X is a local martingale under P and we denote by $\mathcal{M}(X)$ the set of probabilities $Q \ll P$ such that X is a Q local martingale. $\mathcal{M}^e(X)$ is the collection of martingale measures Q (i.e. $Q \in \mathcal{M}(X)$) equivalent to P. Following J.M. Harrison and S.R. Pliska (1981) we say that B $(\in \mathcal{F})$ is a contingent claim if $B \geq 0$. Moreover B is said to be Q-attainable for some $Q \in \mathcal{M}(X)$ if there exists a V such that :

i) $V_t = V_0 + \int_0^t H_s^* dX_s$ where H is integrable w.r.t. X

ii) V is a Q-martingale

iii) $V_1 = B$.

2. MAIN RESULTS

J.M. Harrison and S.R. Pliska (1981) conjectured that the following two statements are true. First, a bounded claim is attainable with one choice of martingale measure iff it is attainable with any choice of equivalent martingale measure ; second, if a claim is attainable under two different choices of martingale measures, then it has the same associated price under each. The answers are positive.

Theorem 2.1 **(C. Stricker, 1990)**

Let B be a bounded random variable, R and Q two equivalent martingale measures. If B is R-attainable, then it is Q-attainable.

The second statement is an easy consequence of the next theorem.

Theorem 2.2 **(J.P. Ansel and C. Stricker, 1992)**

Let X be a local martingale and H a predictable process. Suppose that H is integrable w.r.t. X and $H^*.X$ is bounded below by some constant x. Then $H^*.X$ is a local martingale.

Now assume that B is Q-attainable for some $Q \in \mathcal{M}^e(X)$. Then there exist $x \in \mathbb{R}$ and a predictable process H such that $\forall\, t \in [0,1]$ $E^Q[B \mid \mathcal{F}_t] = x + (H^*.X)_t \geq 0$. Theorem 2.2 states that $H^*.X$ is a nonnegative R-local martingale whenever $R \in \mathcal{M}(X)$. Hence $H^*.X$ is a R supermartingale and $E^R[B] \leq x = E^Q[B]$. Moreover $H^*.X$ is a R martingale iff $E^R[B] = x = E^Q[B]$. So we get the following corollary.

Corollary 2.3

Let R and Q be two equivalent martingale measures such that the contingent claim B is both R and Q-attainable. Then $\forall\, t \in [0,1]$ $E^R[B \mid \mathcal{F}_t] = E^Q[B \mid \mathcal{F}_t]$. In particular $E^R[B] = E^Q[B]$ and $H^R = H^Q$.

Remark 2.4

M. Emery (1981) gives an example of a local martingale X and a predictable process H such that $H^*.X$ is not a local martingale.

We are now in a position to state the main result of this paper which was independently proved recently by S.D. Jacka (1992) and J.P.

Ansel & C. Stricker (1992). It generalizes Theorem 8.5 in Karatzas et al. (1991).

Theorem 2.5

Let B be a contingent claim and $Q \in \mathcal{M}^e(X)$. The following conditions are equivalent :

i) B is Q-attainable

ii) $\forall\, R \in \mathcal{M}^e(X) \qquad E^R[B] \leq E^Q[B] < +\infty$.

The inequality $E^R[B] \leq E^Q[B]$ may be strict for some $R \in \mathcal{M}^e(X)$ (see W. Schachermayer (1992) for a beautiful example). The proof of i) $\Longrightarrow$ ii) is just a straightforward application of theorem 2.2. To establish the reverse implication, we consider the space $\mathcal{L}^1(X \cup \{1\})$, the closure in $H^1(Q)$ of the collection of elements of $H^1(Q)$ which can be written as linear combinations of stochastic integrals w.r.t. elements of $X \cup \{1\}$. By localisation we may assume that $B \in H^1(Q)$. Then we follow the proof of theorem 11.2 of Jacod (1979). Suppose, c is a continuous linear functional on $H^1(Q)$, which vanishes on $\mathcal{L}^1(X \cup \{1\})$. Since the dual of $H^1(Q)$ is $BMO(Q) \subseteq H^\infty_{\text{loc}}(Q)$, there exists $Z \in H^\infty_{\text{loc}}(Q)$ such that $c(M) = E^Q[Z, M]_\infty$ for all $M \in H^1(Q)$. By a new localisation we may assume that $Z \in H^\infty(Q)$ with $\|Z\|_\infty = a$. Now define R by $\frac{dR}{dQ} := 1 + \frac{Z_\infty}{2a}$. Then $R \in \mathcal{M}^e(Q)$ and hence $E[Z_\infty B] \leq 0$. Therefore $E^Q[Z_\infty B] = 0$ and $B \in \mathcal{L}^1(X \cup \{1\})$.

Let us conclude this paper with a nice application of theorem 2.5 given by F. Delbaen and W. Schachermayer (1992).

We assume X is locally bounded. Then it is easy to show that $\mathcal{M}(P)$ is closed in $L^1(P)$. The theorem of Bishop-Phelps states that the set $\{B \in L^\infty : \exists\, R \in \mathcal{M}(P)$ satisfying $E^R[B] \geq E^Q[B]$ for all $Q \in \mathcal{M}(P)\}$ is dense in L^∞. Finally let us recall a deep result of M. Yor (1978) : the set of Q-attainable random variables is $\sigma(L^1, L^\infty)$ closed. F. Delbaen and W. Schachermayer combine these two theorems to get the following result.

Theorem 2.6

If X is locally bounded, then $\mathcal{M}^e(P) = \mathcal{M}(P)$ iff $\mathcal{M}(P) = \{P\}$.

REFERENCES

Ansel, J.P., Stricker, C. (1992) Couverture des actifs contingents. To appear.

Black F., Scholes, M. (1973) The pricing of options and corporate liabilities. *Journal of Political Economy,* 81, 637-654.

Chatelain, M., Stricker, C. (1992) On componentwise and vector stochastic integration. To appear.

Delbaen, F., Schachermayer, W. (1992) A general version of the fundamental theorem of asset pricing. In preparation.

Emery, M. (1981) Compensation de processus à variation finie non localement intégrables. *In Séminaire de Probabilités XIV,* edited by J. Azéma and M. Yor, pp. 152-160. Heidelberg: Springer.

Harrison, J.M., Pliska, S.R. (1981) Martingales and arbitrage on multiperiod securities markets. *Stochastic Processes and Applications,* 11, 215-260.

Jacka, S.D. (1992) A martingale representation result and an application to incomplete financial markets. To appear in *Journal of Mathematical Finance.*

Jacod, J. (1979) *Calcul stochastique et problème de martingales.* Lecture Notes in Mathematics, 714. Springer.

Karatzas, I., Lehoczky, J.P., Shreve, S.E. & Xu, G.L. (1991) Martingale and duality methods for utility maximization in an incomplete market. *SIAM J. Control Optim.,* 29, 702-730.

Merton, R. (1973) The Theory of Rational Option Pricing. *Bell Journal of Economics and Management Science,* 4, 141-183.

Schachermayer, W. (1992) A counter-example to several problems in mathematical finance. To appear.

Stricker, C. (1990) Arbitrage et lois de martingales, *Annales de l'Institut Henri Poincaré,* 26, 451-460.

Yor, M. (1978) Sous-espaces denses dans L^1 ou H^1 et représentation des martingales. *In Séminaire de Probabilités XII,* edited by C. Dellacherie, P.A. Meyer and M. Weil, pp. 265-309. Heidelberg : Springer.

ON THE JOINT DISTRIBUTION OF THE SUPREMUM AND TERMINAL VALUE OF A UNIFORMLY INTEGRABLE MARTINGALE

PIERRE VALLOIS
Laboratoire de Probabilités. Université Pierre et Marie Curie, URA 224, CNRS. Tour 56, 4 place Jussieu. 75252 PARIS CEDEX 05. FRANCE .

INTRODUCTION

1°) The aim of this paper is the study of the joint law of the terminal value and the supremum of a uniformly integrable martingale vanishing at zero. Let us start with the main results.
Let $\mathcal{M}$ (resp. $\mathcal{M}_c$) be the set of uniformly integrable (resp.u.i. and continuous) martingales $(M(t)\ ;\ t \geq 0)$ such that $M(0) = 0$. If M belongs to $\mathcal{M}$, $(\bar{M}(t)\ ;\ t \geq 0)$ denotes its maximum process, and is defined by : $\bar{M}(t)= \sup\{M(u)\ ;\ 0\leq u\leq t\}$, $t\geq 0$.
We know that $\{P(M_\infty \in \cdot),\ M \in \mathcal{M}\}$ (resp. $\{P(M_\infty \in .),\ M \in \mathcal{M}_c\}$) is exactly the set $\mathcal{P}$ of probability measures μ on $\mathbb{R}$, such that $\int |t|\, d\mu(t) < +\infty$ and $\int t\, d\mu(t) = 0$ (see for instance the solutions of the Skorokhod problem given by [1], [2], [4], [10], [15], [16]).
Let M be an element of $\mathcal{M}$, we set $\mu = P(M_\infty \in \cdot)$. Dubins and Gilat [5] proved that there exists a probability measure μ_* on $\mathbb{R}_+$ such that,

$$P(\bar{M}_\infty \geq \lambda) \leq \mu_*[\lambda,+\infty[, \qquad \forall\lambda \geq 0, \tag{1}$$

μ_* is the maximal distribution associated with μ.
Kertz and Rösler [7] characterized the laws of M_∞ and $\bar{M}_\infty$, where M belongs to $\mathcal{M}$: let ν be a probability measure on $\mathbb{R}_+$ and μ an element of $\mathcal{P}$, then there exists $M \in \mathcal{M}$ such that,

$$\nu = P(\bar{M}_\infty \in \cdot) \quad \text{and} \quad \mu = P(M_\infty \in \cdot)$$

if and only if,

$$\mu[\lambda,+\infty[\ \leq \nu[\lambda,+\infty[\ \leq \mu_*[\lambda,+\infty[\ , \quad \forall\lambda \geq 0.$$

2°) Here, we are interested by the law of $\bar{M}_\infty$, when M belongs to $\mathcal{M}$. We start with the continuous case.
Let ν be a probability measure on $\mathbb{R}_+$, a_ν the supremum of the support of ν and

$$\nu(dx) = n(x)dx + \nu_s(dx), \tag{2}$$

the Lebesgue decomposition of ν, ν_s being the singular part of ν with respect to the Lebesgue measure.

Theorem 1 : ([17]) *There exists an element* M *belonging to* $\mathcal{M}_c$, *such that* $\nu = P(\bar{M}_\infty \in \cdot)$, *if and only if,*

$$n \quad \textit{is a.e. positive on} \quad [0,a_\nu[\tag{3}$$

$$\lim_{x\to+\infty} x\,\nu[x,+\infty[\;=\; 0 \tag{4}$$

$$\int_0^{+\infty} |xn(x) - \nu[x,+\infty[\,|\,dx + \int_0^{+\infty} x\nu_s(dx) < +\infty. \tag{5}$$

Let η and η' be two probability measures on $\mathbb{R}_+$. We say that $\eta \le \eta'$ iff

$$\eta[\lambda,+\infty[\;\le\; \eta'[\lambda,+\infty[\quad \text{for every } \lambda \ge 0 .$$

We denote by $\mathcal{S}_{ac}$ the set of probability measures ν on $\mathbb{R}_+$ such that ν is absolutely continuous with respect to the Lebesgue measure on $\mathbb{R}_+$ (i.e. $\nu_s = 0$) and $\nu = P(\bar{M}_\infty \in \cdot)$, for some $M \in \mathcal{M}_c$.

Theorem 2 : *Let ν be a probability measure on $\mathbb{R}_+$. There exists an element* M *belonging to* $\mathcal{M}$ *such that* $\nu = P(\bar{M}_\infty \in \cdot)$, *if and only if, there exists $\nu' \in \mathcal{S}_{ac}$ such that $\nu \le \nu'$.*

3°) Recently Rogers [12] gave a complete answer, describing the joint law of $(\bar{M}_\infty, M_\infty)$, M belonging to $\mathcal{M}$ or $\mathcal{M}_c$. Let η be a probability measure on $\{(x,y) \in \mathbb{R}_+ \times \mathbb{R} \;;\; x \ge y\}$. We introduce the "barycentre" function of η :

$$b(\eta)(t) = \begin{cases} \dfrac{\int y \, 1_{\{x \geq t\}} \, \eta(dx,dy)}{\eta([t,+\infty[\times \mathbb{R})}, & \text{if } \eta([t,+\infty[\times \mathbb{R}) > 0 \\ t \quad \text{if } \eta([t,+\infty[\times \mathbb{R}) = 0 . \end{cases}$$

We can notice that if $\eta = P(\bar{M}_\infty \in \cdot, M_\infty \in \cdot)$ then $b(\eta)(t) = E[M_\infty | \bar{M}_\infty \geq t]$, $t \geq 0$.
We begin with the continuous case.

Theorem 3 ([12]) : *There exists* M *belonging to* $\mathcal{M}_c$, *such that:* $\eta = P(\bar{M}_\infty \in \cdot, M_\infty \in \cdot)$, *if and only if,*

$$\int |y| \, \eta(\mathbb{R}_+, dy) < +\infty \quad , \quad \int y \, \eta(\mathbb{R}_+, dy) = 0 \tag{6}$$

$$b(\eta)(t) = t \qquad \forall t > 0. \tag{7}$$

The general case is stated as follows: $\eta = P(\bar{M}_\infty \in \cdot, M_\infty \in \cdot)$ for some $M \in \mathcal{M}$ if and only if (6) holds and

$$b(\eta)(t) \geq t \qquad \forall t \geq 0 \tag{8}$$

$$b(\eta) \text{ is an increasing function.} \tag{9}$$

We detail the organization of this paper : a first paragraph is devoted to the proof of Theorem 1, and the study of the law of $\bar{M}_\infty$, $M \in \mathcal{M}_c$. In a second paragraph we give a new proof of the result of Rogers [12], (Theorem 3), based on Rost results. We prove Theorem 2 in section 3 .

1. THE LAW OF THE SUPREMUM OF A CONTINUOUS AND U.I. MARTINGALE

Let M be an element of $\mathcal{M}_c$. The key identity is the well known "maximal equality" ,

$$xP(\bar{M}_\infty \geq x) = E(M_\infty 1_{\{\bar{M}_\infty \geq x\}}), \qquad \forall x \geq 0. \tag{1.1}$$

We can notice that (1.1) is equivalent to $b(\eta)(x) = x$, where η is the law of $(\bar{M}_\infty , M_\infty)$.

Let φ be a bounded Borel function defined on $\mathbb{R}_+$ and $\Phi(x) = \int_0^x \varphi(t)\, dt$; $x \geq 0$. We set, $N_t = (\overline{M}_t - M_t)\varphi(\overline{M}_t) - \Phi(\overline{M}_t)$; $t \geq 0$.

Lemma 1: *N is a local martingale. Moreover if φ has a compact support, N is a u.i. martingale, in particular,*

$$E[(\overline{M}_\infty - M_\infty)\varphi(\overline{M}_\infty)] = E[\Phi(\overline{M}_\infty)] . \tag{1.2}$$

Proof: 1) Assume that φ is C^1. By the Ito formula, we get,

$$dN_t = (\overline{M}_t - M_t)\, \varphi'(\overline{M}_t)\, d\overline{M}_t - \varphi(\overline{M}_t)\, dM_t .$$

If t belongs to the support of $d\overline{M}_.$, then $\overline{M}_t - M_t = 0$. This shows that N is a local martingale.
2) If φ has a compact support then the processes $(\overline{M}_t\varphi(\overline{M}_t);\ t \geq 0)$, $(\varphi(\overline{M}_t);\ t \geq 0)$ and $(\Phi(\overline{M}_t)\ ;\ t \geq 0)$ are bounded. Since M is u.i. then $(M_t\varphi(\overline{M}_t);\ t \geq 0)$ is u.i.. We have proved that N is a u.i. process, (1.2) follows immediately. Using the monotone class theorem we get the result. □

Remarks : 1°) Result (1.2) can be found in [1], Corollaire 2.2.
2°) If we choose $\varphi(u) = 1_{\{u>x\}}$, an easy computation shows that (1.2) reduces to (1.1) (see [11], exercise 4.16, p 247). Then (1.1) is equivalent to (1.2).

Proof of Theorem 1 :

1°) We begin by showing the necessary part.
Since $E(|M_\infty|) < +\infty$, by (1.1) it is clear that (4) is realized.
We set $\nu = P(\overline{M}_\infty \in dx)$, $\lambda(x) = E(M_\infty | \overline{M}_\infty = x)$ and $\Gamma = \{x \geq 0\ ;\ \lambda(x) = x\}$.
Let φ be a bounded Borel function with compact support. By (1.2), we obtain,

$$E[(\overline{M}_\infty - \lambda(\overline{M}_\infty))\varphi(\overline{M}_\infty)] = \int_0^\infty P(\overline{M}_\infty \geq t)\ \varphi(t)\ dt.$$

Then,

$$\nu[t,+\infty[dt = (t-\lambda(t))d\nu(t). \tag{1.3}$$

By (2), we get,

$$\nu[t,+\infty[\; = (t-\lambda(t))\, n(t) \qquad \text{a.e.} \quad (\text{on } \Gamma^c), \tag{1.4}$$

$$\operatorname{supp}(\nu_s) \subset \Gamma \tag{1.5}$$

$$\text{the Lebesgue measure of } \Gamma \cap [0,a_\nu] \text{ is zero.} \tag{1.6}$$

Clearly (1.4) implies (3) and (5) is now a direct consequence of (1.4) , (1.5), (1.6) and

$$E[\,|E[M_\infty|\bar{M}_\infty]|\,] \le E[\,|M_\infty|\,] < +\infty.$$

2°) We suppose now that ν is a fixed probability on $\mathbb{R}_+$, with decomposition (2) and satisfying (3), (4) and (5). We set,

$$\lambda(x) = \begin{cases} x - \dfrac{\nu[x,+\infty[}{n(x)} & \text{if } n(x) > 0 \\ x & \text{if } n(x) = 0 \end{cases}$$

$$\beta(x) = \int_0^x \frac{n(u)}{\nu[u,+\infty[}\, du \,, \qquad 0 \le x < a_\nu \,.$$

<u>Lemma 2</u> : *For every* x *in* $[0,a_\nu[$, we have

$$\beta(x) < +\infty, \tag{1.7}$$

$$\int_{[0,x[} \exp(\beta(u))\, \nu_s(du) + \nu[x,+\infty[\, \exp \beta(x) = 1. \tag{1.8}$$

<u>Proof of Lemma 2</u> : Since $\nu[u,+\infty[\; \ge \int_u^{+\infty} n(t)dt$, we get,

$$\beta(x) \le -\ln\left(\int_x^{+\infty} n(t)dt\right) < +\infty \quad , \text{ if } x < a_\nu \,.$$

Let φ be the function defined by,

$$\varphi(x) = \int_{[0,x[} \exp(\beta(u))\, \nu_s(du) + \nu[x,+\infty[\, \exp \beta(x).$$

We differentiate,

$$d\varphi(x) = \exp(\beta(x)) \left[\nu_s(dx) + \frac{n(x)}{\nu[x,+\infty[}\, \nu[x,+\infty[dx - \nu(dx)\right] = 0.$$

Therefore φ is constant, moreover $\varphi(0) = 1$; then (1.8) is satisfied. □

We recall the following result which can be found in ([17], Proposition 3) or in ([17], Lemme 1), and will be useful in the proof of Proposition 1 .

Lemma 3 : *Let* $(M_t \; ; \; t \geq 0)$ *be a martingale such that* $\lim_{t \to +\infty} M_t = M_\infty$ *exists a.s.,* $E[|M_\infty|] < +\infty$, $E[M_\infty] = 0$ *and* $(M_t^- \; ; \; t \geq 0)$ *is u.i. Then* $(M_t \; ; \; t \geq 0)$ *is a u.i. martingale.*

Let $(B_t; \; t \geq 0)$ be a one-dimensional Brownian motion started at 0 and $(S_t \; ; \; t \geq 0)$ its maximum process, i.e, $S_t = \sup_{0 \leq u \leq t} B_u; \; t \geq 0$.
Let $T(x)$ be the stopping time $T(x) = \inf\{t \geq 0, \; B_t = x\}$.
The proof of Theorem 1 will be complete if we establish

Proposition 1 : *Let* A *be a positive r.v. which takes its values in in* $[0,+\infty]$, *such that,*

$$P(A \in dx, \; A < +\infty) = \exp(\beta(x))\nu_s(dx). \tag{1.9}$$

We assume that A *is independent of* $(B_t; \; t \geq 0)$ *and we define* T *and* T' *by* $T' = \inf\{t \geq 0 \; ; \; B_t \leq \lambda(S_t)\}$ *and* $T = T' \wedge T(A)$. *Then* $T < +\infty$ *a.s.,* $(B_{T \wedge t}; \; t \geq 0)$ *is a u.i. martingale and*

$$P(S_T \in dx) = \nu(dx).$$

Proof of Proposition 1 : 1°) By (1.8) we have,

$$\int_{[0,a_\nu[} \exp(\beta(u)) \; \nu_s(du) \leq 1, \text{ and } \int_{[0,a_\nu]} \exp(\beta(u)) \; \nu_s(du) = 1$$

if $a_\nu < +\infty$.
This shows that we can define a r.v. A such that (1.9) holds.
We denote by $(L_t \; ; \; t \geq 0)$ the local time process at 0 of B.
We recall the identity of P. Lévy: $(S-B,S) \overset{(d)}{=} (|B|,L)$ and a result of Jeulin and Yor [6]:
Let h be a positive Borel function, defined on $[0,\delta[$, where $\delta \in]0,+\infty]$, such that,

$$\int_0^x \frac{du}{h(u)} < +\infty \quad , \quad \forall x \in [0,\delta[\ , \tag{1.10}$$

Let U be the following Brownian stopping time,

$$U = \inf\{t \geq 0 \ ; \ |B_t| \geq h(L_t) \quad \text{or} \quad L_t \geq \delta\} \tag{1.11}$$

By (4.1) of ([6]), we have,

$$U < \infty \quad \text{a.s. if } \delta < \infty \quad \text{or } \delta = +\infty \text{ and } \int_0^{+\infty} \frac{du}{h(u)} = +\infty \tag{1.12i}$$

$$L_U \leq \delta \quad \text{and} \quad |B_U| = h(L_U) \quad \text{if} \quad U < \infty. \tag{1.12ii}$$

$$P(L_U \geq x) = \exp - \int_0^x \frac{du}{h(u)} \quad , \quad 0 \leq x \leq \delta. \tag{1.13}$$

2°) Therefore $(T', B_{T'} \ , \ S_{T'}) \overset{(d)}{=} (U, L_U - |B_U|, L_U)$, where U is defined by (1.11) and associated with $h(x) = x - \lambda(x)$ and $\delta = a_\nu$. By the previous step and Lemma 2, we obtain,

$$T' < +\infty \quad \text{if } a_\nu < +\infty \quad \text{or} \quad a_\nu = +\infty \quad \text{and} \quad \beta(+\infty) = +\infty,$$

$$A < +\infty \quad \text{if } a_\nu < +\infty \quad \text{or} \quad a_\nu = +\infty \quad \text{and} \quad \beta(+\infty) < +\infty.$$

Moreover by (1.13), we get,

$$P(S_{T'} \geq x) = \exp - \beta(x) \qquad 0 \leq x \leq a_\nu \ . \tag{1.14}$$

Let f be a positive Borel function and $\Delta = E[f(S_T)]$. Then,

$$\Delta = [f(S_{T'})\, 1_{\{S_{T'} \leq A\}}] + E[f(A)\, 1_{\{A < S_{T'}\}}].$$

By (1.8) and (1.9), we have,

$$P(A \geq x) = \nu[x, +\infty[\ \exp \beta(x). \tag{1.15}$$

Recall that A is independent of $(B_t \ ; \ t \geq 0)$, an easy calculation shows,

$$\Delta = \int_0^{+\infty} f(x)\, n(x)\, dx + \int_0^{+\infty} f(x)\, \nu_s(dx) = \int_0^{+\infty} f(x)\, \nu(dx). \tag{1.16}$$

This means that the law of S_T is ν.

3°) It remains to show that $(B_{T\wedge t}\ ;\ t \geq 0)$ is a u.i. martingale. As before, we have successively,

$$E[|B_T|] = E[|\lambda(S_{T'})| 1_{\{S_{T'} \leq A\}}] + E[A\ 1_{\{S_{T'} > A\}}],$$

$$E[|B_T|] = \int_0^{+\infty} |\lambda(x)| n(x)dx + \int_0^{+\infty} x\ \nu_s(dx).$$

Assumption (5) implies that, $E[|B_T|] < +\infty$.

By a similar calculation, we get,

$$E[B_T] = \int_0^{+\infty} (xn(x) - \nu[x,+\infty[)dx + \int_0^{+\infty} x\ d\nu_s(x).$$

Using (5) once more, we obtain,

$$\int_0^{+\infty} (xn(x) - \nu[x,+\infty[)dx = \lim_{n\to+\infty} \int_0^{n} (xn(x) - \nu[x,+\infty[)dx.$$

Since $\int_0^{n} \nu[x,+\infty[dx = n\ \nu[n,+\infty[+ \int_0^{n} x\ d\nu(x)$, and (4) holds, we have: $E[B_T] = 0$.

By lemma 3, it is sufficient to show that $(B^-_{T\wedge t}\ ;\ t \geq 0)$ is a u.i. process. We introduce an increasing family of stopping times $(T_n\ ;\ n \geq 1)$, by setting,

$$T'_n = \inf\{t \geq 0,\ B_t \leq \lambda_n(S_t)\},\ T_n = T'_n \wedge T(A),$$

where

$$\lambda_n = \lambda \vee (-n).$$

Let $M > 0$, and Δ_n defined by,

$$\Delta_n = E(B^-_{T_n} 1_{\{B^-_{T_n} \geq M\}}).$$

We remark that $B^-_{T_n} \leq n$ and $B^-_{T_n} = (\lambda_n(S_{T'_n}))^- 1_{\{S_{T'_n} \leq A\}}$; then

$$\Delta_n = \int_0^{a_\nu} \frac{(\lambda_n(x))^-}{x-\lambda_n(x)} P(S_{T'_n} \geq x) P(A \geq x) 1_{\{\lambda_n(x)^- \geq M\}} dx.$$

Since $\lambda \leq \lambda_n$, $T'_n \leq T'$, therefore by (1.14) and (1.15) we obtain,

$$\Delta_n \leq \int_0^{a_\nu} \frac{-\lambda_n(x)}{x-\lambda_n(x)} \nu[x,+\infty[\; 1_{\{\lambda_n(x) \leq -M\}} dx.$$

If $\lambda(x) \leq -n$, then $\frac{-\lambda_n(x)}{x-\lambda_n(x)} = \frac{n}{x+n} \leq \frac{-\lambda(x)}{x-\lambda(x)}$, and

$$\Delta_n \leq \int_0^{a_\nu} \frac{-\lambda(x)}{x-\lambda(x)} \nu[x,+\infty[\; 1_{\{\lambda(x) \leq -M\}} dx$$

$$\Delta_n \leq \int_0^{a_\nu} (-xn(x) + \nu[x,+\infty[) \; 1_{\{\lambda(x) \leq -M\}} dx.$$

It is now clear that, uniformly with respect to n, Δ_n goes to 0, as $M \longrightarrow +\infty$. This means that the sequence of r.v. $(B^-_{T_n} \; ; \; n \geq 1)$ is u.i..

Moreover for large n, $T_n = T$, then $B^-_{T_n}$ converges a.s. to B^-_T .

By Neveu ([9], Proposition II.5.4), $B^-_{T_n}$ converges to B^-_{T}, in L^1.

Let $(\mathcal{F}_t \; ; \; t \geq 0)$ be the natural filtration of the Brownian motion B. Since $(B^-_{t\wedge T_n} \; ; \; t \geq 0)$ is a submartingale bounded by n , we have,

$$B^-_{t\wedge T_n} \leq E(B^-_{T_n} | \mathcal{F}_t) \quad , \forall \; t \geq 0 \; .$$

When n goes to infinity, we get: $B^-_{t\wedge T} \leq E(B^-_T | \mathcal{F}_t)$ for every $t \geq 0$.

The last inequality implies that $(B^-_{t\wedge T} \; ; \; t \geq 0)$ is u.i. . □

Let ν be a probability measure on $\mathbb{R}_+$, with decomposition (2). We set : $\nu_a(dx) = n(x)dx$, ν_a is the absolutely continuous part of ν. The assumptions (3) and (4) are clear. We are now dealing with (5). It is easy to show that (5) is equivalent to,

$$\int_0^{+\infty} \left| xn(x) - \nu_a[x,+\infty[\right| dx < +\infty, \tag{1.16}$$

$$\int_0^{+\infty} x\, \nu_s(dx) < +\infty. \tag{1.17}$$

Then $\nu = \nu_a + \nu_s$, satisfies (5) iff ν_s admits a first moment and ν_a satisfies (1.16).

We remark that if ν admits a first moment, then (5) holds. The converse can be false (see Lemma 4) . In ([17], Proposition 4) there is an example of a probability measure ν such that : $\nu = \nu_a$, $a_\nu = +\infty$, verifying (3), (4) but not (5).

Lemma 4 : *We set* $\varepsilon(x) = x\,\nu_a[x,+\infty[$. *We assume that* ε *is a decreasing function on* $[\alpha,+\infty[$ *for some* $\alpha > 0$ *and* $\lim_{x\to+\infty} \varepsilon(x) = 0$. *Then* (1.16) *holds.*

Proof : We have, $\varepsilon'(x) = \nu_a[x,+\infty[- x\,n(x)$. Then,

$$\int_\alpha^{+\infty} |\varepsilon'(x)|dx = -\int_\alpha^{+\infty} \varepsilon'(x)dx = \varepsilon(\alpha) < +\infty. \qquad \square$$

We denote by $\mathcal{S}_{ac}$ the set of probability measures ν on $\mathbb{R}_+$ verifying (3), (4), (5) and such that $\nu(dx) = n(x)\,dx$ (i.e. $\nu = \nu_a$).

If f belongs to $L^1([0,1])$, $H(f)$ denotes the Hardy transform of f, recall that,

$$H(f)(t) = \frac{1}{1-t}\int_1^t f(u)du \quad , \qquad 0 \le t < 1.$$

Let $\mathcal{H}$ be the set of functions f belonging to $L^1([0,1])$, such that $\int_0^1 f(u)du = 0$ and $H(f)$ is an increasing function.

Proposition 2 ([17]) : *Let* ν *be a probability measure on* $\mathbb{R}_+$. *Then* ν *belongs to* $\mathcal{S}_{ac}$ *iff there exists a function* f *in* $\mathcal{H}$ *such that* ν *is the image by* $H(f)$ *of the Lebesgue measure on* $[0,1]$. *If* ν *belongs to* $\mathcal{S}_{ac}$ *, we have,*

$$f(x) = k(x) - \frac{1-x}{n(k(x))}, \qquad 0 < x < 1,$$

where n *is the density of* ν *and* k *is the inverse of* $x \longrightarrow \nu[0,x]$.

Proof : We just sketch the proof.

We remark that ν is the image by $H(f)$ of the Lebesgue measure on $[0,1]$ iff,

$$\nu[0,x] = \int_0^1 1_{\{t\ ;\ H(f)(t) \le x\}}\, dt, \qquad x \ge 0. \tag{1.18}$$

We use the fact that $H(f)$ is increasing, then (1.18) is equivalent to ,

$$k(x) = H(f)(x), \quad 0 < x < 1. \tag{1.19}$$

We differentiate (1.19) we obtain,

$$f(x) = k(x) - (1-x)\, k'(x). \tag{1.20}$$

It is easy to check that f belongs to $L^1([0,1])$ iff ν satisfies (1.16).

Since $H(f)'(x) = \dfrac{H(f)(x) - f(x)}{1-x}$, $H(f)$ is increasing iff $H(f) \ge f$.

By (1.19) and (1.20), we have,

$$H(f)(x) - f(x) = k(x) - f(x) = (1-x)k'(x) \ge 0. \qquad \square$$

2. A DIRECT PROOF OF THEOREM 3

Let M be an element of $\mathcal{M}_c$ and $\eta = P(\bar{M}_\infty \in \cdot, M_\infty \in \cdot)$. It is clear that η satisfies (6) and (7) ((7) is exactly equivalent to (1.1)). We check now that these conditions are also sufficient. Our proof is based on the general results of Rost [14] , concerning the stopping distributions of a Markov process. As in paragraph 1, $(B(t)\ ;\ t \ge 0)$ denotes a one-dimensional Brownian motion started at 0. We first remark that,

$$(S_{u+v}\, , B_{u+v}) = (S_u \vee (\max_{u \le w \le v} B_w), B_{u+v}) \tag{2.1}$$

for every $u \ge 0$, $v \ge 0$.

Then $X = (S,B)$ is a Markov additive process ([3]) which takes its values in $\mathbb{R}_+ \times \mathbb{R}$.

Rost [14] proved that if X is a Markov process on a L.C.D. space

$(E,\mathcal{E})$ and λ_1, λ_2 are two probability measures on $(E,\mathcal{E})$ such that $\lambda_1 U$ is σ-finite, where U is the potential of X, then

there exists a stopping time T such that $P_{\lambda_1}(X_T \in \cdot) = \lambda_2$ **(2.2)**

if and only if,

$$\lambda_1 U(f) \geq \lambda_2 U(f) \quad , \quad \forall f \geq 0 \quad , \quad f \;\; \mathcal{E}\text{-measurable.} \tag{2.3}$$

Lemma 5 : *For every positive Borel function we have,*

$$E\left[\int_0^{+\infty} g(S_t,B_t)dt\right] = 2\int g(u,v)\, 1_{\{u>0,u>v\}}du\, dv.$$

Proof : The law of (S_t,B_t) is well known (see for instance [8], théorème 4.2.2, p. 195),

$$P(S_t \in du, B_t \in dv) = 2\,\varphi_t(2u-v)\, 1_{\{u>0,u>v\}}du\, dv,$$

where $\varphi_t(x) = (2\pi t^3)^{-1/2}\, x \exp - (x^2/2t)$.
We set,

$$\Delta = \int_0^{+\infty} E[g(S_t,B_t)]\, dt.$$

Then,

$$\Delta = 2\int g(u,v)(2u-v)\, h(u,v)\, 1_{\{u>0,u>0\}}du\, dv,$$

where

$$h(u,v) = \int_0^{+\infty} (2\pi t^3)^{-1/2} \exp -\left(\frac{(2u-v)^2}{2t}\right) dt.$$

We set $s = \dfrac{(2u-v)^2}{2t}$, we obtain ,

$$h(u,v) = \frac{1}{\sqrt{\pi}}\,\frac{1}{2u-v}\int_0^{+\infty} s^{-1/2}\, e^{-s} ds = \frac{1}{2u-v}\,. \qquad \square$$

Lemma 6 : *Let* $x > 0$, $x \geq y$ *and* f *be a positive Borel function, then* ,

$$\frac{1}{2} U(x,y,f) = (x-y)\, \varphi(x) - \Phi(x) - \int_{-\infty}^{x} f(x,v)(v-y)^{+}\, dv$$

$$+ \int f(u,v) 1_{\{u>0,u>v\}}\, du\, dv \qquad \textbf{(2.4)}$$

where $\varphi(x) = \int_{-\infty}^{x} f(x,v)dv$ *and* $\Phi(x) = \int_{0}^{x} \varphi(u)du$.

Proof : Let $(P_t\ ;\ t \geq 0)$ be the semi-group of (S,B). By (2.1), we have, $P_t(x,y,f) = E[f(x \vee (S_t+y),\ B_t+y)]$.

By definition, $U(x,y,f) = \int_{0}^{+\infty} P_t(x,y,f)dt$.

We use lemma 5, we obtain,

$$U(x,y,f) = 2 \int f(x \vee (s+y), t+y)\ 1_{\{s>0,s>t\}} ds\ dt.$$

We set $u = s+y$ and $v = t+y$, $U(x,y,f)$ splits into two parts,

$$U(x,y,f) = U_1(x,y,f) + U_2(x,y,f)$$

where

$$U_1(x,y,f) = 2 \int f(x,v) 1_{\{v\leq x\}} (x - v \vee y)\ dv,$$

$$U_2(x,y,f) = 2 \int f(u,v)\ 1_{\{u\geq x,u>v\}}\ du\ dv.$$

Since $x - v \vee y = x-y - (v-y)^{+}$ then

$$U_1(x,y,f) = 2(x-y) \int f(x,v)\ 1_{\{v\leq x\}} dv - 2 \int f(x,v)\ 1_{\{v\leq x\}}\ (v-y)^{+} dv. \ \square$$

Remark : Let ψ_1 be the function defined by ,

$$\psi_1(x,y) = - \int_{-\infty}^{x} f(x,v)(v-y)^{+}\ dv.$$

Lemma 1 implies that $((S_t-B_t)\ \varphi(S_t) - \Phi(S_t)\ ;\ t \geq 0)$ is a local martingale. Since $\frac{\partial\psi_1}{\partial x}(x,x) = 0$ and $\frac{\partial^2\psi_1}{\partial y^2}(x,y) = -2f(x,y)$, then $(\psi_1(S_t,B_t)\ ;\ t \geq 0)$ is a local supermartingale. By (2.4) we check that $(U(S_t,B_t,f)\ ;\ t \geq 0)$ is a supermartingale ; a result which is well known.

From now on we suppose that η is a probability measure on $\{(x,y)\ ;\ x \geq 0 \text{ and } x \geq y\}$.

Lemma 7 : *If* η *satisfies* (6) *and* (7) *then* ,

$$\delta_0 U(f) \geq \eta U(f) \text{ for every } \mathcal{B}(\mathbb{R}_+ \times \mathbb{R})\text{-measurable } f \geq 0, \quad \mathbf{(2.5)}$$

where δ_0 *denotes the Dirac measure at* $(0,0)$.

Proof : Let f be a positiveand $\mathcal{B}(\mathbb{R}_+\times\mathbb{R})$-measurable function. By (2.4), (2.5) is equivalent to,

$$\int\Big\{(x-y)\ \varphi(x) - \Phi(x) - \int_{-\infty}^{x} f(x,v)(v-y)^+dv\Big\}\eta(dx,dy) \leq 0. \qquad \mathbf{(2.6)}$$

Recall that η satisfies

$$\int y\ 1_{\{x\geq t\}}\ \eta(dx,dy) = t\ \eta([t,+\infty[\ \times\ \mathbb{R}) \quad \forall t \geq 0.$$

We use (6), we obtain ,

$$\int(x-y)\ 1_{[0,t]}(x)\ \eta(dx,dy) = \int x\ 1_{[0,t]}(x)\ \eta(dx,dy) + \int y\ 1_{\{x\geq t\}}\ \eta(dx,dy)$$

$$= \int\Big(\int_0^x 1_{[0,t]}(u)du\Big)\eta(dx,dy).$$

By the monotone class theorem, we get

$$\int(x-y)\ \varphi_0(x)\ \eta(dx,dy) = \int\Big(\int_0^x \varphi_0(u)du\Big)\eta(dx,dy).$$

for every positive Borel function φ_0
If we choose $\varphi_0 = \varphi$, we see that (2.6) reduces to,

$$- \int\left(\int_{-\infty}^{x} f(x,v)\ (v-y)^{+}dv\right)\ \eta(dx,dy) \le 0.$$

It ic obviouc that thic inequality holds, since $f > 0$. □

We can apply the result of Rost [14] : there exists a finite stopping time T such that $P(S_T \in \cdot, B_T \in \cdot) = \eta$. The proof of theorem 3 will be finished if we establish ,

Lemma 8 : $(B_{T\wedge t}\ ;\ t \ge 0)$ *is a u.i. martingale.*

Proof : For every $n \ge 1$, let σ_n be the stopping time,

$$\sigma_n = \inf\{t \ge 0\ ;\ B_t \ge n\}.$$

We set $T_n = T \wedge \sigma_n$ and $\Delta_n = E[|B^+_{T_n} - B^+_T|]$.

We have,

$$\Delta_n = E[|n - B^+_T| 1_{\{S_T \ge n\}}] = \int |n-y^+|\ 1_{\{x\ge n\}}\eta(dx,dy).$$

Then

$$\Delta_n \le n\ \eta([n,+\infty[\ \times \mathbb{R}) + \int |y|\ 1_{\{x\ge n\}}\ \eta(dx,dy).$$

By (6) and (7), we obtain that $\lim_{n\to+\infty} \Delta_n = 0$.

In other words, $B^+_{T_n}$ converges in L^1 to B^+_T. Moreover $T_n = T$ for large n, therefore,

$$B^+_{t\wedge T} \le E(B^+_T | \mathcal{F}_{t\wedge T}),\ t \ge 0.$$

This shows that $(B^+_{t\wedge T}\ ;\ t \ge 0)$ is a u.i. process.

We set $M_t = -B_{t\wedge T}$, $t \ge 0$. Using once more (6) and lemma 3, we prove that $(B_{T\wedge t}\ ;\ t \ge 0)$ is a u.i. martingale. □

3. PROOF OF THEOREM 2

Let ν be a probability measure on $\mathbb{R}_+$ such that $\nu = P(\bar{M}_\infty \in \cdot)$, $M \in \mathcal{M}$. By inequality (1) , there exists $\mu \in \mathcal{P}$ such that : $\nu \le \mu_*$. The Azéma and Yor theorem ([1],théorème 3.4), tells us that μ_*

belongs to $\mathcal{S}_{ac}$.

We now establish the converse. We assume that $\nu \le \nu'$,for some $\nu' \in \mathcal{S}_{ac}$. Then there exists a continuous and uniformly integrable martingale M', started at 0 such that $P(\bar{M}_\infty \in \cdot) = \nu'$. We denote by η' the joint law of $(\bar{M}'_\infty, M'_\infty)$.

Since $\nu]\lambda,+\infty[\;\le\; \nu']\lambda+\infty[$, for every $\lambda \ge 0$, there exists a right-continuous and increasing function c such that,

$$\nu]\lambda,+\infty[\;=\; \nu']c(\lambda),+\infty[\,, \qquad \forall\lambda \ge 0 \tag{3.1}$$

$$c(\lambda) \ge \lambda \qquad , \quad \forall\lambda \ge 0. \tag{3.2}$$

Let η be the image of η' by the map $(x,y) \longrightarrow (\varphi(x),y)$, where φ is the right-inverse of c.

Since $E(|M'_\infty|) < +\infty$ and $E(M'_\infty) = 0$, then

$$\int |y| \; \eta'(\mathbb{R}_+,dy) = \int |y| \; \eta(\mathbb{R}_+,dy) < +\infty,$$

$$\int y \; \eta'(\mathbb{R}_+,dy) = \int y \; \eta(\mathbb{R}_+,dy) = 0.$$

By definition, for every $\lambda \ge 0$, we have,

$$\eta(]\lambda,+\infty[\times \mathbb{R}) = \int 1_{\{\varphi(x)>\lambda\}} \; \nu'(dx).$$

If $\varphi(c(\lambda)) = \lambda$ (resp. $\varphi(c(\lambda)) > \lambda$), $x > c(\lambda) \Longleftrightarrow \varphi(x) > \lambda$ (resp. $x \ge c(\lambda) \Longleftrightarrow \varphi(x) > \lambda$).

Recall that ν' doest not charge the points, it follows that,

$$\eta(]\lambda,+\infty[\times \mathbb{R}) = \nu']c(\lambda),+\infty[\;=\; \nu]\lambda,+\infty[.$$

By the same way, we get,

$$\int 1_{\{x>\lambda\}} \; y \; \eta(dx,dy) = \int 1_{\{x>c(\lambda)\}} \; \eta'(dx,dy).$$

But the right hand-side is equal to $\nu']c(\lambda),+\infty[$.

Therefore, $b(\eta)(\lambda) = c(\lambda)$; $\forall\lambda \ge 0$.

We can apply the second theorem of Rogers, η satisfies (6),(8) and (9), there exists $M \in \mathcal{M}$ such that $\nu = P(\bar{M}_\infty \in \cdot)$. □

REFERENCES

1. Azéma , J. , Yor , M. (1979) (i) Une solution simple au problème de Skorokhod. (ii)Le problème de Skorokhod ; compléments à l'exposé précédent. *InSéminaire de Probabilités XIII,* Vol 721, pp 90-117 ,pp 625- 633 Lect. Notes Math. Berlin

Heidelberg New York : Springer .

2. Chacon , R. V. , Walsh J. B. (1976) One dimensional potential embedding .*In Séminaire de Probabilités X.* Vol 511, pp 19-23, Lect. Notes Math., Berlin Heidelberg New York: Springer.
3. Cinlar E. (1972) Markov additive process I . *Z. Warscheinlichkeitstheorie verw. Geb.* , 24 , 85-93 .
4. Dubins L. E. (1968) On a theorem of Skorohod . *The Annals of Math. Statistics,* Vol 39 , N°6 , 2094-2097 .
5. Dubins L. E. , Gilat D. (1978) On the distribution of maxima of martingales. *Proc. of the A.M.S.,* vol. 68, N° 3 ,337-338. .
6. Jeulin T. , Yor M. (1981) Sur les distributions de certaines fonctionnelles du mouvement brownien. *In Séminaire de Probabilités XV.* Vol 850 , pp 210-226 , Lect. Notes Math ,Berlin Heidelberg New York : Springer .
7. Kertz R.P. , Rösler U. (1990) Martingales with given maxima and terminal distributions. *Israël Journal of Math.* Vol. 69, N° 2, 173-192.
8. Lévy P. (1965) *Processus stochastiques et mouvement brownien.* Gauthier Villars .
9. Neveu J. (1970) *Bases mathématiques du calcul des Probabilités.* Masson.
10. Perkins E. (1986) The Cereteli-Davis solution to the H^1-embedding problem and an optimal embedding in Brownian motion .*In Seminar on Stochastic Processes,* pp 172-223 ,Birkhaüser, Boston .
11. Revuz D., Yor M. (1991) *Continuous martingales and brownian motion.* Springer Verlag.
12. Rogers L. C. G. (1992) The joint law of the maximum and terminal value of a martingale. Preprint.
13. Root D. H. (1969) The existence of certain stopping times on Brownian motion. *The Annals of Math. Stat.,*Vol 40,N°2, 715-718.
14. Rost H. (1971) The stopping distributions of a Markov Process. *Inventiones math.* 14, 1-16 .
15. Skorokhod A. (1965) *Studies in the theory of random process .* Addison-Wesley. Reading Mass.
16. Vallois P. (1983) Le problème de Skorokhod sur $\mathbb{R}$: une approche avec le temps local . *In Séminaire de Probabilités XVII* Vol. 986 , pp 227-239 ,Lectures Notes Math , Berlin Heidelberg New York : Springer .
17. Vallois P. (1991) Sur la loi du maximum et du temps local d'une martingale continue uniformément intégrable.Preprint.
18. Yor M. (1991) Sur les décompositions affines d'une variable stable d'indice 1/2 . *Prépublication N°79 du Laboratoire de Probabilités de l'Université de Paris VI .*

TIME REVERSAL FOR MARKOV PROCESSES WITH LÉVY MEASURE

MATTHIAS WEBER
Department of Mathematics, Technical University of Dresden, Mommsenstr.13, O-8027 Dresden, Germany

Abstract The infinitesimal generator of the time reversal of a Markov process with Lévy measure having a singularity is given using an approximation of its resolvent.

INTRODUCTION

One dimensional Markov processes X given by an infinitesimal generator of the form

$$(Af)(x) = (D_m D_s f)(x) + \int_0^1 (f(y)-f(x)) \frac{n_x(dy)}{|s(y)-s(x)|} \quad (1)$$

$f \in D(A) \subset C[0,1]$, with boundary conditions were considered by Langer and Schenk[1]. The generator consists of a diffusion part $D_m D_s$, the generalized second order derivative introduced by Feller and Krein, and an integral part with a Lévy measure N:

$$N(x,dy) = \frac{n_x(dy)}{|s(y)-s(x)|}, x,y \in [0,1], \ x \neq y,$$

describing possible jumps of the process. We mention that the notation of the diffusion part used by Langer and Schenk[1] can easily be transformed into the notation used here. It was shown that the resolvent operators R_λ of A can be approximated by the resolvents $R_{\lambda,\varepsilon}$, $\lambda>0$, $0<\varepsilon<1$, of infinitesimal generators A^ε which are defined by equation (1) if we replace n_x by n_x^ε:

$$n_x^\varepsilon(dy) = I_{[\varepsilon,1]}(|x-y|)\ n_x(dy). \tag{2}$$

Here the function I is the indicator function. Evidently the singularity in the integral part of A does not appear in A^ε. Langer and Schenk[1] gave a representation of R_λ, $\lambda>0$, as an integral operator with a continuous kernel and they showed that its natural extension $\tilde{R}_\lambda$ to the space $L^2(m)$ is the resolvent of an infinitesimal generator $\tilde{A}$ in $L^2(m)$. Also a representation of its adjoint was given. However, there is no probabilistic interpretation of this adjoint.

Time reversal for Markov processes with an infinitesimal generator of a little more general form as in relation (1) but without a singularity in the integral part was considered by Weber[2] and the infinitesimal generator of the time reversed process was given.

In this note we combine these results to approximate the adjoint resolvent $\tilde{R}_\lambda^*$ of $\tilde{R}_\lambda$ in the space $L^2(M_\upsilon)$ by resolvents $\tilde{R}_{\lambda,\varepsilon}^*$ which are the adjoints of $\tilde{R}_{\lambda,\varepsilon}$ in $L^2(M_{\upsilon,\varepsilon})$. Here the $\tilde{R}_{\lambda,\varepsilon}$ are the extensions of the resolvents of the time reversal of the process X^ε, given by A^ε, to

$L^2(M_{\upsilon,\varepsilon})$. M_υ and $M_{\upsilon,\varepsilon}$ are the potential measures of X and X^ε which depend on the initial distribution υ. Using the properties of the density g_υ of M_υ with respect to the measure m we show that the restriction of $\tilde{R}^*_\lambda$ to $C[0,1]$ is the resolvent of a Feller process $\hat{X}$ which is the time reversal of X. An explicit representation of the infinitesimal generator of $\hat{X}$ is given. We mention that the approximation in $L^2(M_\upsilon)$ is also possible in the case of a higher singularity in the integral part of the generator. However it seems to be more difficult to obtain a Feller process associated with the approximated resolvent in this case.

1 THE GENERATORS IN C[0,1]

Let s and m be continuous strictly increasing functions on [0,1] and n_x, $x\in[0,1]$, a family of measures on [0,1] with the following properties:

(i) n_x depends *-weakly continuous on x,

(ii) $n_x(\{x\}) = 0$, $x\in[0,1]$.

In the space C[0,1] we consider the operators B and B^ε defined on $D(D_mD_s)$:

$$(Bf)(x) := \int_0^1 (f(y)-f(x))\frac{n_x(dy)}{|s(y)-s(x)|}, \quad (3)$$

$$(B^\varepsilon f)(x) := \int_0^1 (f(y)-f(x))\frac{n^\varepsilon_x(dy)}{|s(y)-s(x)|}. \quad (4)$$

Here the measures n^ε_x, $0<\varepsilon<1$, are defined by the relation (2). The restrictions A and A^ε of

$$D_m D_s + B \text{ and } D_m D_s + B^{\varepsilon} \tag{5}$$

by boundary conditions

$$\Phi_i(f) = \kappa_i f(i) + (-1)^{i+1}\pi_i (D_s f)(i) = 0, \tag{6}$$

$i=0,1$, $\kappa_i \geq 0$, $\pi_i > 0$, are infinitesimal generators of strongly continuous nonnegative contraction (s.c.n.c.) semigroups in $C[0,1]$. Their resolvents R_λ, $R_{\lambda,\varepsilon}$, $\lambda>0$, admit an integral representation:

$$(R_\lambda f)(x) = \int_0^1 G(x,y;\lambda) f(y) dm(y), \tag{7}$$

$x \in [0,1]$, $f \in C[0,1]$, and the corresponding representation for $R_{\lambda,\varepsilon}$ with continuous strictly positive kernels G and G_ε (comp. Langer and Schenk[1]). The restriction of $D_m D_s$ by (6) we denote by A^0 and $R_\lambda^0 = (\lambda - A^0)^{-1}$, $\lambda>0$. In the following we assume:

$$\kappa_0 + \kappa_1 > 0, \tag{8}$$

such that the representation (7) holds for $\lambda=0$ too. Langer and Schenk[1] have shown that

$$\| R_{\lambda,\varepsilon} - R_\lambda \| \to 0 \text{ as } \varepsilon \downarrow 0. \tag{9}$$

Integrating the equation $\lambda R_{\lambda,\varepsilon} f - A^{\varepsilon} R_{\lambda,\varepsilon} f = f$ over $[0,x]$ it is not hard to find a function $C(\lambda)$ such that:

$$|(D_s R_{\lambda,\varepsilon} f)(x)| \leq C(\lambda) \| f \| \ , \lambda>0. \tag{10}$$

Let X and X^{ε} be the Feller processes given by A and A^{ε} with a fixed initial distribution υ. Further M_υ and $M_{\upsilon,\varepsilon}$ denote the potential measures of these processes:

$$M_\upsilon(\Gamma) = \mathbb{E}_\upsilon \int_0^\infty I_\Gamma(X_t)dt = \int_\Gamma g_\upsilon(x)dm(x), \Gamma\in\mathcal{B}[0,1], \quad (11)$$

i.e. $g_\upsilon(x)=\int_0^1 G(y,x;0)d\upsilon(y)$ is the potential density (similar relations hold for $M_{\upsilon,\varepsilon}$ and $g_{\upsilon,\varepsilon}$). It is shown by Weber[2] that $g_{\upsilon,\varepsilon}$ can be chosen continuous and strictly positive and that the time reversal $\hat{X}^\varepsilon_t = X^\varepsilon_{(L-t)-}$, $0<\varepsilon<1$, $t>0$, with respect to a cooptional time L is given by the infinitesimal generator $\hat{A}^\varepsilon$:

$$(\hat{A}^\varepsilon \hat{f})(x) = \frac{d(D_{\hat{s}}\hat{f})(x) - \hat{f}(x)d\hat{k}(x) + \int_{y=0}^1 (\hat{f}(y)-\hat{f}(x))\hat{\mu}^\varepsilon(dy\times dx)}{d\hat{m}(x)}, \quad (12)$$

$$\hat{\Phi}_i(\hat{f}) = \hat{\kappa}_i \hat{f}(i) + (-1)^{i+1}\hat{\pi}_i (D_{\hat{s}}\hat{f})(i) = 0, \ i=0,1, \text{ where:}$$

$$d\hat{s}(x) = g_{\upsilon,\varepsilon}(x)^{-2}ds(x), \quad d\hat{m}(x) = g_{\upsilon,\varepsilon}(x)^2 dm(x),$$

$$d\hat{k}(x) = g_{\upsilon,\varepsilon}(x)\left(d\upsilon(x) - \sum_{i=0}^{1} \upsilon(\{i\})d\varepsilon_i(x)\right),$$

$$\hat{\kappa}_i = \frac{\pi_i}{g_{\upsilon,\varepsilon}(i)}\upsilon(\{i\}), \quad \hat{\pi}_i = g^{-2}_{\upsilon,\varepsilon}(i)\pi_i,$$

$$\hat{\mu}(dy\times dx) = \frac{g_{\upsilon,\varepsilon}(x)\ n^\varepsilon_y(dx)dm(y)}{g_{\upsilon,\varepsilon}(y)\,|s(x)-s(y)|}, x,y\in[0,1],\ x\neq y.$$

Here ε_i denotes the degenerated distribution in i. An essential tool to compute $\hat{A}^\varepsilon$ is the following duality relation holding for all $f,g\in C[0,1]$:

$$\int_0^1 f(R_{\lambda,\varepsilon}g)dM_{\upsilon,\varepsilon} = \int_0^1 g(\hat{R}_{\lambda,\varepsilon}f)dM_{\upsilon,\varepsilon}. \quad (13)$$

In the next section we need:

<u>Lemma 1.</u> It holds $\|g_{\upsilon,\varepsilon} - g_\upsilon\| \to 0$ as $\varepsilon\downarrow 0$.

Proof. The assumption (8) is equivalent to $\lambda\|R_\lambda\|<1$. As the kernels G and G_ε are positive this implies:

$$1>\lambda\|R_\lambda\|=\sup_{x\in[0,1]}\lambda(R_\lambda 1)(x)=\sup_{x\in[0,1]}\lim_{\varepsilon\downarrow 0}\lambda(R_{\lambda,\varepsilon}1)(x).$$

As $R_{\lambda,\varepsilon}1$ converges in $C[0,1]$ there exist a c, $0<c<1$, and an ε_0, $0<\varepsilon_0<1$, such that $\lambda\|R_{\lambda,\varepsilon}\|<1-c$ for all $0<\varepsilon<\varepsilon_0$. Using $R_{0,\varepsilon}=(1-\lambda R_{\lambda,\varepsilon})^{-1}R_{\lambda,\varepsilon}$ we get

$$\|R_{0,\varepsilon}\|\le\lambda\|R_{\lambda,\varepsilon}\|\|(1-\lambda R_{\lambda,\varepsilon})^{-1}\|$$
$$\le\|\sum_{i=0}^{\infty}(\lambda R_{\lambda,\varepsilon})^i\|\le\sum_{i=0}^{\infty}(1-c)^i=c^{-1}.$$

Further we have $R_\lambda=R^0_\lambda+R_\lambda BR^0_\lambda$, $\lambda\ge 0$, and BR^0_0 is an integral operator with a kernel $B(x,y)$ and measure dm (see Langer and Schenk[1]). So we get

$$g_\upsilon(y)=\int_0^1\left(G_0(x,y;0)+\int_0^1 G(x,z;0)B(z,y)dm(z)\right)d\upsilon(x)$$

$$=\int_0^1\left(G_0(x,y;0)+(R_0B(\cdot,y))(x)\right)d\upsilon(x)$$

and a similar relation for $g_{\upsilon,\varepsilon}$. Here $G_0(\cdot,\cdot;\lambda)$ denotes the kernel of R^0_λ. We obtain

$$\|g_{\upsilon,\varepsilon}-g_\upsilon\|=\sup_{y\in[0,1]}\left|\int_0^1\left((R_0B(\cdot,y))(x)-(R_{0,\varepsilon}B(\cdot,y))(x)\right.\right.$$
$$\left.\left.+(R_{0,\varepsilon}(B(\cdot,y)-B_\varepsilon(\cdot,y)))(x)\right)d\upsilon(x)\right| \tag{14}$$

$$\le\sup_{y,z\in[0,1]}\left(\left|\int_0^1((R_0-R_{0,\varepsilon})B(\cdot,y))(x)d\upsilon(x)\right|\right.$$
$$\left.+\frac{1}{c}\,|B(z,y)-B_\varepsilon(z,y)|\right).$$

With $R_{0,\varepsilon}-R_0=R_{0,\varepsilon}(B_\varepsilon-B)R_\lambda(1-\lambda R_\lambda)$ and relation (10) it follows that $R_{0,\varepsilon}f\to R_0f$ as $\varepsilon\downarrow 0$ for all $f\in B[0,1]$.

Thus, the first term of the right hand side of relation (14) vanishes as $\varepsilon\downarrow 0$. The properties of the kernels B_ε and B imply the same for the other term. ∎

This convergence and the fact that g_υ is continuous and strictly positive imply the existence of constants $\varepsilon_o>0$, $c>0$ and C such that

$$c<g_{\upsilon,\varepsilon}(x)<C, \tag{15}$$

for all $x\in[0,1]$ and $\varepsilon<\varepsilon_o$. Probabilistic interpretations of all parts appearing in A^ε and $\hat{A}^\varepsilon$ are given by Weber[2].

2 THE GENERATORS IN L^2

Following the considerations of Langer and Schenk[1] we denote by $\tilde{R}_{\lambda,\varepsilon}$ and $\tilde{R}_\lambda$, $\lambda\geq 0$, $0<\varepsilon<1$, the extensions of the resolvents $R_{\lambda,\varepsilon}$ and R_λ to the spaces $L^2(M_{\upsilon,\varepsilon})$ and $L^2(M_\upsilon)$ according to equation (7). It is not hard to check that $\tilde{R}_{\lambda,\varepsilon}$ and $\tilde{R}_\lambda$ are resolvents of infinitesimal generators of s.c.n.c. semigroups in the respective L^2-spaces. Here the contraction property follows from the sub-invariance of the potential measures. Further the extension $\tilde{R}^*_{\lambda,\varepsilon}$ of $\hat{R}_{\lambda,\varepsilon}$ to $L^2(M_{\upsilon,\varepsilon})$ is the adjoint of $\tilde{R}_{\lambda,\varepsilon}$. Combining relations (7) and (13) it follows:

$$(\tilde{R}^*_{\lambda,\varepsilon}f)(x) = \int_0^1 \hat{G}_\varepsilon(x,y;\lambda)f(y)dm(y) \tag{16}$$

$$= \int_0^1 G_\varepsilon(y,x;\lambda)f(y)\frac{g_{\upsilon,\varepsilon}(y)}{g_{\upsilon,\varepsilon}(x)}\,dm(y),$$

$x\in[0,1]$, $f\in L^2(M_{\upsilon,\varepsilon})$, $\lambda\geq 0$, $0<\varepsilon<1$. By approximation the relation (13) can be generalized to:

$$\langle f, \tilde{R}_{\lambda,\varepsilon} g\rangle_{\varepsilon} = \langle \tilde{R}^{*}_{\lambda,\varepsilon} f, g\rangle_{\varepsilon} \tag{17}$$

where $\langle .,. \rangle_{\varepsilon}$ denotes the inner product in this spaces. From Lemma 1 and relation (9) it follows:

$$\langle \tilde{R}^{*}_{\lambda,\varepsilon} f, g\rangle_{\varepsilon} = \langle f, \tilde{R}_{\lambda,\varepsilon} g\rangle_{\varepsilon} \to \langle f, \tilde{R}_{\lambda} g\rangle, \text{ as } \varepsilon\downarrow 0, \tag{18}$$

such that $\tilde{R}^{*}_{\lambda,\varepsilon} f$ tends to a function $\tilde{R}^{*}_{\lambda} f$ for each $f \in L^2(M_{\upsilon})$ and $\lambda>0$. We remark that all $L^2(M_{\upsilon,\varepsilon})$, $L^2(M_{\upsilon})$ and $L^2(m)$ contain the same functions (comp. (15)). Now it is easy to check:

<u>Proposition 2.</u> $\tilde{R}^{*}_{\lambda}$, $\lambda\geq 0$, is the resolvent of an infinitesimal generator $\tilde{A}^{*}$ of a s.c.n.c. semigroup in $L^2(M_{\upsilon})$.

Let us now compare the adjoint $\tilde{A}^{*}$ with the $L^2(m)$ adjoint given by Langer and Schenk[1] which we denote here by $\tilde{A}^{*}_{m}$ with resolvent $\tilde{R}^{*m}_{\lambda}$. Let $f,g \in L^2(m)$. Then:

$$\int_0^1 f(\tilde{R}_{\lambda} g) g_{\upsilon} dm = \int_0^1 g(\tilde{R}^{*m}_{\lambda} f g_{\upsilon}) dm = \int_0^1 g(\tilde{R}^{*}_{\lambda} f) g_{\upsilon} dm. \tag{19}$$

Thus, $h \in D(\tilde{A}^{*})$ iff $(hg_{\upsilon}) \in D(\tilde{A}^{*}_{m})$ and $\tilde{A}^{*} h = g_{\upsilon}^{-1} \tilde{A}^{*}_{m}(hg_{\upsilon})$. This transformation is similar to the superharmonic transformation in Nagasawa's theorem.

3 THE RESTRICTION TO C[0,1]

According to (16), by $\hat{R}_{\lambda}$ let us denote the restriction of $\tilde{R}^{*}_{\lambda}$ to C[0,1]. Using (15) similar considerations as above (comp. (10)) lead to a function $\hat{C}(\lambda)$ such that:

$$|(D_s\hat{R}_{\lambda,\varepsilon}f)(x)|\leq\hat{C}(\lambda)\|f\|, \tag{20}$$

$\lambda>0$, $0<\varepsilon<1$, $f\in C[0,1]$. With $\lambda\|\hat{R}_{\lambda,\varepsilon}\|\leq 1$ we get from the Arzela-Ascoli-theorem that for fixed $\lambda>0$ and $f\in C[0,1]$ the set $\{\hat{R}_{\lambda,\varepsilon}f,\ 0<\varepsilon<1\}$ is compact in $C[0,1]$. Now, with (18) $\hat{R}_{\lambda,\varepsilon}f$ tends to $\hat{R}_\lambda f$ in $C[0,1]$ for all $\lambda>0$, $f\in C[0,1]$ as $\varepsilon\downarrow 0$. Observing the representation (12) of $\hat{A}^\varepsilon$ and relation (20) it follows that $\hat{R}_\lambda f$ is a solution of the boundary value problem:

$$\lambda h(x)d\hat{m}(x) - d(D_{\hat{s}}h)(x) + h(x)d\hat{k}(x) \tag{21}$$

$$- \int_0^1(h(y)-h(x))\frac{\hat{\mu}(dy\times dx)}{|s(y)-s(x)|} = f(x)d\hat{m}(x),\ x\in(0,1),$$

$$\hat{\Phi}_i(h)=\hat{\kappa}_i h(i)+(-1)^{i+1}\hat{\pi}_i(D_{\hat{s}}h)(i)=0,\ i=0,1,\ \text{where:}$$

$$d\hat{s}(x) = g_\upsilon(x)^{-2}ds(x),\ d\hat{m}(x) = g_\upsilon(x)^2dm(x),$$

$$d\hat{k}(x) = g_\upsilon(x)\Big(\ d\upsilon(x) - \sum_{i=0}^{1}\upsilon(\{i\})d\varepsilon_i(x)\Big), \tag{22}$$

$$\hat{\kappa}_i = \frac{\pi_i}{g_\upsilon(i)}\upsilon(\{i\}),\qquad \hat{\pi}_i = g_\upsilon^{-2}(i)\pi_i,$$

$$\hat{\mu}(dy\times dx) = \frac{g_\upsilon(x)}{g_\upsilon(y)}n_y(dx)dm(y),\ x,y\in[0,1],\ x\neq y.$$

Using the local maximum principle for generalized second order derivatives it is possible to show that the solution of (21) is unique. Further properties of $\hat{R}_\lambda$, $\lambda\geq 0$, such as the resolvent equation, nonnegativity and the contraction property follow evidently from the respective properties of the $\hat{R}_{\lambda,\varepsilon}$ via convergence.

Theorem 3. $\hat{R}_\lambda$ is the resolvent of an infinitesimal generator $\hat{A}=\lambda-\hat{R}_\lambda^{-1}$ of a s.c.n.c. semigroup in C[0,1] and the associated Feller process is the time reversal (with respect to any cooptional time) of the Feller process associated with the generator A.

Proof. To apply the Hilla-Yoside-theorem it remains to show that for at least one $\lambda>0$ the set $\{\hat{R}_\lambda f,\ f\in C[0,1]\}$, i.e. the domain of $\hat{A}$, is dense in C[0,1]. By the Hahn-Banach-theorem this follows if we show that for any function Π of bounded variation the relation

$$\int_0^1 \hat{R}_\lambda f \, d\Pi = 0 \text{ for all } f\in C[0,1], \tag{23}$$

implies $\Pi=0$. Equation (23) gives

$$0 = \int_0^1\int_0^1 G(y,x;\lambda)g_\upsilon(y)g_\upsilon^{-1}(x)f(y)dm(y)d\Pi(x)$$

for all $f\in C[0,1]$ (comp. (16)) such that

$$\int_0^1 G(y,x;\lambda)g_\upsilon^{-1}(x)d\Pi(x) = 0 \text{ for m-almost all y.}$$

As $0<g_\upsilon<\infty$ it suffices to show that $d\bar{\Pi}=g_\upsilon^{-1}d\Pi\equiv 0$. From the representation of $R_\lambda=R_\lambda^o(1+K_\lambda)$ given by Langer and Schenk[1] where K_λ is an integral operator in C[0,1] we get:

$$0=\int_0^1 G_o(y,x;\lambda)d\bar{\Pi}(x)+\int_0^1\int_0^1 G_o(y,t;\lambda)K_\lambda(t,x)dm(t)d\bar{\Pi}(x)$$

for all $y\in[0,1]$ (the right hand side is continuous). Application of dD_s to this equation yields

$$0 = d\bar{\Pi}(y) + \int_0^1 K_\lambda(y,x)d\bar{\Pi}(x)dm(y),\ y\in[0,1].$$

With $Q(y) = \int_0^1 K_\lambda(y,x)d\overline{\Pi}(x)$ we obtain

$$0=\int_0^1 \hat{R}_\lambda f \, d\Pi = -\int_0^1\int_0^1 G(y,x;\lambda)g_\upsilon(y)f(y)dm(y)Q(x)dm(x).$$

Using the integral representation of $\tilde{R}_\lambda$ this is

$$0 = -\int_0^1 (\tilde{R}_\lambda Q)(y)g_\upsilon(y)f(y)dm(y)$$

for all $f\in C[0,1]$. With $\tilde{R}_\lambda Q\in C[0,1]$ it follows $Q=0$ m-a.e. and the first statement of the theorem is proved.

From equations (9),(13) and Lemma 1 we obtain the duality of X and $\hat{X}$ with respect to the potential measure M_υ via convergence. Thus, X and $\hat{X}$ are time reversals of each other.∎

In virtue of the equations (21) and the transformations (22) the generator $\hat{A}$ can be written in the form

$$(\hat{A}\hat{f})(x)=\frac{d(D_{\hat{s}}\hat{f})(x)-\hat{f}(x)d\hat{k}(x)+\displaystyle\int_0^1 \frac{\hat{f}(y)-\hat{f}(x)}{|s(y)-s(x)|}\hat{\mu}(dy\times dx)}{d\hat{m}(x)}.$$

(Substitute $\hat{R}_\lambda f=h$ in (21) and recall $\lambda\hat{R}_\lambda f-\hat{A}\hat{R}_\lambda f=f$.)

REFERENCES

1. H.Langer and W.Schenk, A class of infinitesimal generators of one-dimensional Markov processes, III. Transition probabilities, *Math.Nachr.*, **102** (1981), 25-44
2. M.Weber A class of infinitesimal generators and time reversal for the corresponding one-dimensional Markov processes, *Math. Nachr.* **157** (1992), 65-75

Index